SpringerBriefs in Applied Sciences and Technology

Mathematical Methods

Series editor

Anna Marciniak-Czochra, Heidelberg, Germany

For further volumes:
http://www.springer.com/series/11219

Hamid Reza Noori

Hysteresis Phenomena in Biology

Hamid Reza Noori
Central Institute for Mental Health
Institute for Psychopharmacology
Mannheim
Germany

ISSN 2191-530X   ISSN 2191-5318   (electronic)
ISBN 978-3-642-38217-8   ISBN 978-3-642-38218-5   (eBook)
DOI 10.1007/978-3-642-38218-5
Springer Heidelberg New York Dordrecht London

Library of Congress Control Number: 2013955060

Printed on acid-free paper

Springer is part of Springer Science+Business Media (www.springer.com)

We should be careful to get out of an experience only the wisdom that is in it—and stop there.

Mark Twain

Preface

For over a century, nonlinear phenomena of hysteresis type have been ubiquitous in different areas of science and technology from the physics of materials to the meta-physics of the human behavior. However, mathematical approaches to investigate biological processes with hysteresis nonlinearities have a much shorter history. Although the idea of modeling chemical and biological systems with multiple steady states and state transition dynamics (switches) goes back to Lotka (1924) and Volterra (1931), it was by the early 1990s that a sufficient number of discovered molecular mechanisms with bistable dynamical behavior gave rise to mathematical investigation of biological switches and hysteresis phenomena and a steady increase in the number of publications from 99 articles in 1990 to 562 in the year 2012 (www.ncbi.nlm.nig.gov/pubmed). To date, mathematical models of hysteresis are used to describe a variety of biological phenomena including the dynamics of metabolic networks, conformational changes of transmembrane proteins, and the biomechanical properties of organs such as the lungs and the eye.

Despite the growing interest in this field of research, the access to its theoretical foundations is limited to expert mathematicians, physicists, and engineers. While the seminal monographs by Krasnosel'skii and Pokrovskii (1989), Visintin (1994), Brokate and Sprekels (1996), and Mayergoyz (2003) enlighten the mathematical theory of hysteresis with a view toward physical and engineering problems, less attention was paid to its applications in biology and chemistry. To close this gap, this monograph is written by an interdisciplinary researcher for graduate students and interdisciplinary researchers. It is an introductory graduate-level textbook on the fundamental concepts of hysteresis with a view toward biological applications, for students and researchers with an interest in mathematical and computational biology who already have a solid acquaintance with dynamical systems, ordinary differential equations, and linear functional analysis.

This book comprises two parts and is mostly based on my lectures on dynamical systems and on mathematical models of hysteresis in the University of Heidelberg. In the first part (Chaps. 1–3), we follow the footsteps of Krasnosel'skii and Pokrovskii (1989) and systematically introduce the mathematics of hysteresis. Chapter 1 is an introduction in the concept of hysteresis in light of biological processes with a particular focus on biological switches and bistable phenomena. In analogy to the spectral decomposition of self-adjoint operators in the functional analysis, all Preisach models are superpositions of elementary nonlinear hysteresis operators (rectangular loops resembling bistable systems) called hysterons. This concept might appear as a very intuitive one for biologists and therefore builds the theoretical basis of this monograph. Thus, Chap. 2 recapitulates the fundamental definitions and ideas from nonlinear analysis on the stability and bifurcation theory of dynamical systems and prepares the transition toward hysteresis models. The global aim of this chapter is to provide the necessary definitions and theorems with an emphasis directed toward hysteresis so that the interested reader can focus on the essential aspects. Particularly, we will recapitulate the Lyapunov stability and Andronov's structural stability concepts, which are also of general interest for mathematical modeling with ordinary differential equations in biology. The information in this chapter is mainly complied from the monographs on the theory of dynamical systems and ordinary differential equations by Verhulst (2008) and Perko (2008) and the experienced reader may skip this chapter. Chapter 3 is primarily concerned with Preisach models of hysteresis, their identification problems, and their numerical simulations. In particular, we will deal with the classical Preisach theory of hysteresis for continuous and periodic input functions, the parallel and sequential superpositions of hysterons, and the impact of perturbations of input function on the systems. Furthermore, in compliance with Mayergoyz (2003) we will discuss a strategy to identify the weights of Preisach operators from experimental data and we discuss an approach using first-order transition curves for numerical implementation of hysteresis models. On the basis of these mathematical theories, the second part of the book (Chap. 4) demonstrates a few examples of applications of hysteresis theory in the modeling and numerical simulation of biological and chemical systems, particularly the development of mathematical models of bistability and hysteresis-like behavior in different biological disciplines.

Based on the large number of relevant publications, no attempt has been made to exhaust the list of references but to introduce those with the most recognizable impact in their fields. This imperfection is reflected throughout this monograph, as I have made no attempt at all to include all available information. Instead, my focus was to utilize the interested reader with a systematic introduction to a very complicated mathematical theory and reveal its potential for modeling biological and chemical processes. My first contact with mathematical models of hysteresis was during my conversations with Professor Willi Jäger as I was a Ph.D. student in his workgroup some years ago. I was fascinated by its nonlinear nature and was ambitious to study and apply it on every problem that I was working on. However, my impression was that the mathematics of hysteresis and its applications that were of interest for me, namely neuroscientific problems, were still disconnected and to use the models for biological problems, one had to go through the entire mathematical theories. My feeling was further corroborated by the feedbacks that I received from the students and postdocs who attended my lecture on mathematical models of hysteresis in the institute for applied mathematics of the University of Heidelberg, which motivated me to rewrite my lecture notes with a constant view toward the applications. I was very fortunate that my old friend Dr. J. P. Schmidt (Springer) approached me to publish those notes as an introductory book. I would also like to express my gratitude to Professor B. Jacobs (Princeton University), Professor Y. Sinai (Princeton University), Professor A. Marciniak-Czochra (University of Heidelberg), and Professor R. Spanagel (Central Institute for Mental Health) for the very fruitful discussions that my research has benefited from.

My research was financially supported by the Bundesministerium für Bildung und Forschung (NGFN Plus; FKZ: 01GS08152, FKZ: 01GS08155; FKZ: 01GS0815, fundings from the Bernstein Center for Computational Neuroscience initiative; FKZ: 01GQ1003B), and the Deutsche Forschungsgemeinschaft (DFG): Reinhart-Koselleck Award SP 383/5-1 that I acknowledge with gratitude.

Mannheim, Spring 2013   Hamid Reza Noori

References

Brokate M, Sprekels J (1996) Hysteresis and phase transitions. Springer, New York

Krasnosel'skii MA, Pokrovskii AV (1989) Systems with hysteresis. Springer, Heidelberg

Lotka AJ (1924) Elements of physical biology. Dover, New York

Mayergoyz ID (2003) Mathematical models of hysteresis and their applications. Elsevier, New York

Perko L (2008) Differential equations and dynamical systems. Springer, Heidelberg

Verhulst F (2008) Nonlinear differential equations and dynamical systems. Springer, Berlin Heidelberg

Visintin A (1994) Differential models of hysteresis. Springer, Berlin

Volterra V (1931) Lecons sur la theorie de la lutte pour la vie. Gauthier-Villars, Paris

Contents

Part I
Mathematical Models of Hysteresis

Chapter 1
Introduction

The last decades have been witnessing the rise of utilizations of mathematical models
and simulations in life sciences. This tendency is mostly supported by the significant progresses in experimental methodologies as well as recent advancements in
computer sciences that lead to noticeable improvements of the quantitative description of biological processes. Consequently, over the years mathematical modeling
has reached the level of maturity to keep pace with the current biological research
and play a complementary role, particularly in situations in which the experimental
limitations such as insufficient spatiotemporal resolution and/or the complexity of
the systems exacerbate further advancements by experimental approaches.

It is becoming increasingly clear that many biological systems are governed by
highly non-linear bi- or multi-stable processes, which may switch between discrete
states, induce oscillatory behavior or define their dynamics based on a functional
relationship with the memory of input stimuli. However, in light of the experimental
difficulties to capture the underlying mechanisms of such non-linear processes, mathematical modeling may become very useful. For years, the concept of hysteresis is
used to describe similar phenomena in magnetism, optics and material sciences etc.
and has recently attracted the attention of scholars to investigate biological switches
and memory-dependent dynamical systems within its framework. From switches
in protein-DNA interactions (Chatterjee et al. 2008), microscopic cellular signaling
pathways with bistable molecular cascades (Angeli et al. 2004; Qiao et al. 2007), cell
division, differentiation, cancer onset and apoptosis (Eissing et al. 2004; Kim et al.
2007; Wilhelm 2009), protein folding (Andrews et al. 2013) and purinergic neuron-
astrocyte interactions in the brain (Noori 2011) up to macroscopic biomechanics of
cornea (Congdon et al. 2006) and lung deformations (Escolar and Escolar 2004),
hysteresis phenomena are ubiquitous in biology.

Since the focus of attention towards hysteresis has mostly been the physical phenomena, most textbooks dealing with this concept have adapted the terminology
and ideas of related fields of research such as control theory and magnetism, which
often makes it difficult to access this theory for computational biologists and other
researchers that are not familiar with the utilized terms. Following Krasnosel'skii

H. R. Noori, *Hysteresis Phenomena in Biology*, SpringerBriefs in Mathematical Methods,
DOI: 10.1007/978-3-642-38218-5_1, © The Author(s) 2014

Fig. 1.1 Hysteresis loop corresponding to a non-linear relationship between input $x(t)$ and output functions $y(t)$. Although many hysteresis phenomena exhibit hysteresis loops, the occurrence of loops should not be regarded as an essential aspect of hysteresis. In fact, one can easily consider rate-independent models in which no loop occurs as in some cases the hysteresis region is even unbounded

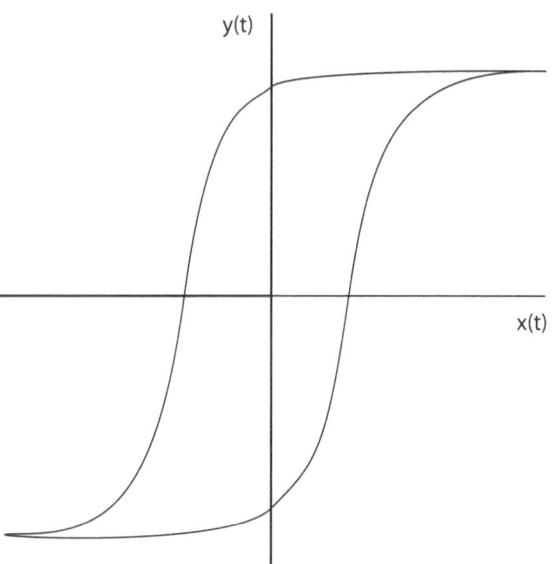

and Pokrovskii (1989), this monograph is intended to provide an introductionary material purely expressed in mathematical terminology and detached applications. This strategy allows the readers to study the theory from their own perspective and discover its power easily for their research. Although an entire chapter is devoted to examples of hysteresis phenomena in biology, the reader will not be involved with the applications and biological terms while studying the mathematical models of hysteresis.

As a result of ages of interdisciplinary work on hysteresis phenomena, a stringent mathematical and universal definition of hysteresis is lacking. Therefore, we begin with a general definition of hysteresis in order to avoid any confusion and ambiguity. Hysteresis, often associated with a hysteresis loop (Fig. 1.1), refers to a non-linear multi-branch operator, which transforms the extreme values of the input functions into branch transitions. In this book, we will only consider rate-independent hysteresis operators, for which the branches of the non-linearities are solely determined by the memory of input extrema and not on the speed of input variations between extreme points. Rate-independent hysteresis operator are commonly categorized into two groups:

- hysteresis operators with local memories. For these operators, the output function $y(t)$ for $t \geq t_0$ is defined by the input function $x(t)$ at $t \geq t_0$ and the initial state of the system at an instance of time t_0; and
- hysteresis operators with non-local memories. In contrast to hysteresis operators with local memories, these operators define their output $y(t)$ for $t \geq t_0$ not only by the initial state of the system at t_0 and the input function $x(t)$ for $t \geq t_0$, but also on past extreme values of the input function $x(t)$.

For hysteresis operators with local memories each reachable point in the (x, y) diagram corresponds to a uniquely defined state of the system, which determines the behavior of the hysteresis operators uniquely for increasing and decreasing inputs $x(t)$. In contrast, at each point of the (x, y) diagram of hysteresis operators with non-local memories, there are infinitely many curves that may define the future behavior of the operator. Hereby, each curve is characterized by a specific pattern of extreme values of the input function $x(t)$. Although the mathematics of hysteresis operators with local memories is extensively studied by differential and algebraic equations, these operators appear to be not in agreement with experimental facts (Mayergoyz 2003). Therefore and in compliance with the other textbooks addressing mathematical models of hysteresis phenomena, we will focus on hysteresis operators with non-local memories in this monograph. For this purpose, we will discuss the basics of scalar Preisach models for continuous monotonic and periodic functions. The intuitive nature of Preisach model of hysteresis, specifically the discontinuous Preisach model that can be represented as a weighted parallel connection of elementary hysteresis operators (Fig. 1.2) the so-called hysterons, makes it an appropriate

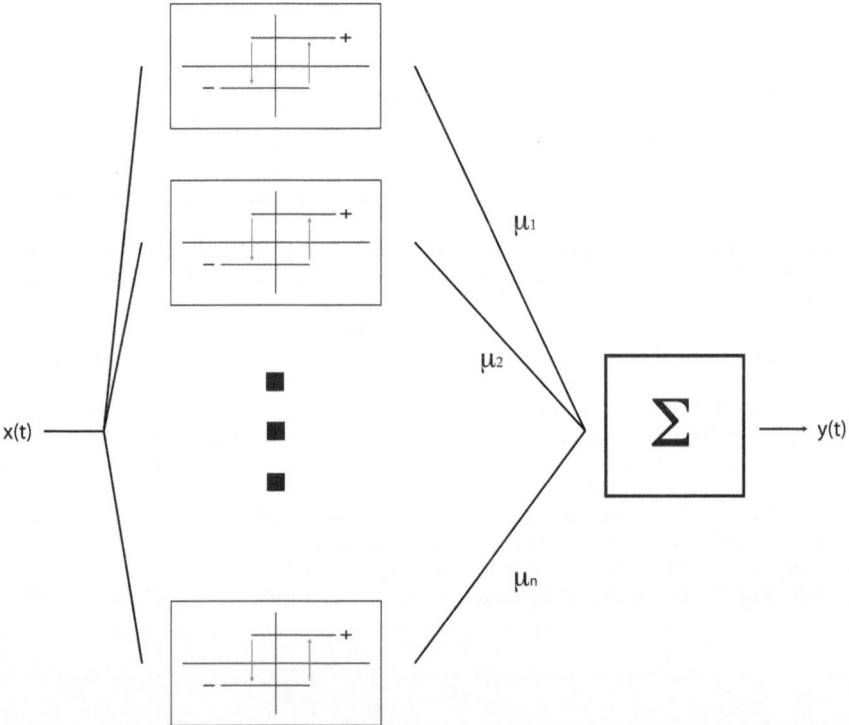

Fig. 1.2 Schematic representation of discontinuous Preisach model as a weighted parallel connection of elementary switch operators (hysterons)

framework for biological applications as hysterons may easily be associated with biological switches and bistable systems.

This formalism not only enables us to formalize and simulate biological switches but it further provide us with the necessary mathematical tools to investigate different compositions of switches and thereby study the interactive mechanisms underlying biological non-linearities. This monograph is only intended as an introduction to the subject and interested readers may also refer to the textbooks by Krasnosel'skii and Pokrovskii (1989); Visintin (1994); Brokate and Sprekels (1996); Mayergoyz (2003). Despite this fact, serious attempt has been made to systematically provide all necessary definitions, theorems and concepts for mathematical modeling of hysteresis phenomena in biology and therefore interdisciplinary researchers and graduate students may benefit from the structured and goal-directed content of this book.

References

Andrews BT, Capraro DT, Sulkowska JI, Onuchic JN, Jennings PA (2013) Hysteresis as a marker for complex, overlapping landscapes in proteins. J Phys Chem Lett 4:180–188

Angeli D, Ferrell JE Jr, Sontag ED (2004) Detection of multistability, bifurcations, and hysteresis in a large class of biological positive-feedback systems. Proc Natl Acad Sci USA 101:1822–1827

Brokate M, Sprekels J (1996) Hysteresis and phase transitions. Springer, New York

Chatterjee A, Kaznessis YN, Hu WS (2008) Tweaking biological switches through a better understanding of bistability behavior. Curr Opin Biotechnol 19:475–481

Congdon NG, Broman AT, Bandeen-Roche K, Grover D, Quigley HA (2006) Central corneal thickness and corneal hysteresis associated with glaucoma damage. Am J Ophthalmol 141: 868–875

Eissing T, Conzelmann H, Gilles ED, Allgoewer F, Bullinger E, Scheurich P (2004) Bistability analyses of a caspase activation model for receptor-induced apoptosis. J Biol Chem 279:36892–36897

Escolar JD, Escolar A (2004) Lung hysteresis: a morphological view. Histol Histopathol 19:159–166

Kim D, Rath O, Kolch W, Cho K-H (2007) A hidden oncogenic positive feedback loop caused by crosstalk between Wnt and ERK pathways. Oncogene 26:4571–4579

Krasnosel'skii MA, Pokrovskii AV (1989) Systems with hysteresis. Springer, Heidelberg

Mayergoyz ID (2003) Mathematical models of hysteresis and their applications. Elsevier, New York

Noori HR (2011) Substantial changes in synaptic firing frequencies induced by glial ATP hysteresis. Biosystems 105(3):238–242

Qiao L, Nachbar RB, Kevrekidis IG, Shvartsman SY (2007) Bistability and oscillations in the Huang-Ferrell model of MAPK signaling. PLoS Comput Biol 3:1819–1826

Visintin A (1994) Differential models of hysteresis. Springer, Berlin

Wilhelm T (2009) The smallest chemical reaction system with bistability. BMC Syst Biol 3:90

Chapter 2
Bifurcation Theory and Bistability

2.1 Basic Definitions and Ideas

In general, a dynamical system is a semigroup G with identity element e acting on a set M, i.e. there exists a mapping

$$T : G \times M \to M$$
$$(g, x) \mapsto T_g(x)$$

such that $T_g \circ T_h = T_{g \circ h}$ and $T_e = \mathbb{I}$. In the following, we shall consider dynamical systems of the form

$$\dot{x} = f(t, x) \tag{2.1}$$

using Newton's notation $\dot{x} = \frac{dx}{dt}$. t is often referred to as the time. $f : G \to \mathbb{R}^n$ is a continuous vector field on G an open subset of $\mathbb{R}^{n+1}$. The choice of a first order dynamical system is not a restriction as one can easily represent the general nth order equation $x^{(n)} = g(t, x, x', x^{(2)}, \ldots, x^{(n-1)})$ with $x^{(n)} := \frac{d^n x}{dt^n}$ by a system of first order equations of the form 2.1. We call a continuously differentiable vector values function on an interval $I \subset \mathbb{R}$ a solution, if it satisfies the Eq. 2.1. The existence and uniqueness of such a solution depend on the properties of the vector field f, particularly its Lipschitz continuity as the next theorem shows.

Theorem 2.1 *Consider the initial value problem*

$$\dot{x} = f(t, x), \quad x(t_0) = x_0$$

with $x \in D = \{x \mid \parallel x - x_0 \parallel \le d\} \subset \mathbb{R}^n$, $|t - t_0| \le a$ and $a, d > 0$ constants. If the vector field $f(t, x)$ is Lipschitz continuous in x and in $M = [t_0 - a, t_0 + a] \times D$, then the initial value problem has one and only one solution $x(t; t_0, x_0)$ for $|t - t_0| \le \inf(a, \frac{d}{\parallel f \parallel})$.

H. R. Noori, *Hysteresis Phenomena in Biology*, SpringerBriefs in Mathematical Methods, 7
DOI: 10.1007/978-3-642-38218-5_2, © The Author(s) 2014

Proof The reader should consult an introductionary textbook on ordinary differential equations (Perko 2008; Verhulst 2008) for the proof. □

Definition 2.1 Consider the autonomous (not explicitly dependent on t) dynamical system $\dot{x} = f(x)$ with $x \in D \subset \mathbb{R}^n$. Then D is called the phase space. A point $(x_1(t), \ldots x_n(t))$ in the phase space for a certain t is then called a phase point. The motion of a set of phase points along the corresponding orbits is called the phase flow.

Definition 2.2 Let $\dot{x}_i = f_i(x)$ for $i = 1, \ldots, n$ be the components of the autonomous dynamical system. Then, the solutions of the system $\frac{dx_j}{dx_i} = \frac{f_j(x)}{f_i(x)}$ for $i, j = 1, \ldots, n, i \neq j$ in the phase space for all $f_i(x) \neq 0$ are called orbits.

The orbits in phase space can take different forms and some can even form singularities and degenerate into a point. Such an orbit is called a critical point.

Definition 2.3 Considering the dynamical system $\dot{x} = f(x)$, a point $a \in \mathbb{R}^n$ with $f_i(a) = 0$ for all i is called a critical or an equilibrium point.

Notice that an equilibrium point corresponds to an equilibrium solution $x(t) = a$ for all time, which can never be reached in finite time.

Definition 2.4 An equilibrium point $x = a$ of the equation $\dot{x} = f(x)$ in $\mathbb{R}^n$ is called a positive attractor, if there exists a neighborhood $N_a \subset \mathbb{R}^n$ of a such that $x(t_0) \in N_a$ implies $\lim_{t \to \infty} x(t) = a$. If an equilibrium point $x = a$ has this property for $t \to -\infty$, then it is called a negative attractor.

Definition 2.5 A cycle or periodic orbit is any closed solution curve in the phase space that is not an equilibrium point.

The easiest way to investigate the qualitative behavior of a dynamical system in terms of the asymptotic properties of solutions and the trajectories is to analyze the behavior of equilibrium points and periodic orbits particularly their stability, which basically means that the solutions do not move away from the equilibrium and periodic solutions if the system is exposed to small perturbations. The concept of stability is also crucial for biological processes as it expresses to robustness of biological systems with respect to variations in their interactions with the environment.

Let us consider $\dot{x} = f(t, x)$ with $(t, x) \in \mathbb{R}^{n+1}$ and the vector field f being continuous in t and x and Lipschitz continuous in x; without loss of generality, let us further assume $x = 0$ is an equilibrium point of the dynamical system such that $f(t, 0) = 0$ for all $t \in \mathbb{R}$.

Definition 2.6 Considering the above mentioned conditions and a neighborhood $N \subset \mathbb{R}^n$ of $x = 0$, then the solution $x = 0$ is called Lyapunov-stable, if for any $\varepsilon > 0$ and starting time t_0 one can find a $\delta(\varepsilon, t_0) > 0$ such that

$$\| x_0 \| \leq \delta \text{ yields } \| x(t; t_0, x_0)) \| \leq \varepsilon \text{ for } t \geq t_0.$$

In a very similar manner, we can define stability for dynamical oscillations or periodic solutions of the dynamical systems:

Definition 2.7 Consider $\dot{x} = f(t, x)$ with a periodic solution $\phi(t)$, then the periodic solution $\phi(t)$ is called Lyapunov-stable, if for any $\varepsilon > 0$ and starting time t_0 there exists a $\delta(\varepsilon, t_0) > 0$ such that

$$\| x_0 - \phi(t_0) \| \le \delta \text{ yields } \| x(t; t_0, x_0)) - \phi(t) \| \le \varepsilon \text{ for } t \ge t_0.$$

Definition 2.8 If a dynamical system $\dot{x} = f(x)$ possesses two (or more) stable equilibrium points, then it is called bistable (or multistable).

The stability of equilibrium solutions or of periodic solutions can be studied often by local analysis of the linearized system. In other words, neglecting the higher order terms of the extension of the dynamical system $\dot{x} = \frac{\partial f}{\partial x}(a)(x - a) + \mathcal{O}(x^2)$ in a neighborhood of these special solutions, we obtain a linear equation $\dot{x} = Ax$ with A an $n \times n$ matrix that is the subject of stability analysis. This strategy has been justified more than a century ago by Poincare and Lyapunov and will be the subject of the next section of this chapter.

2.2 Stability of Linear Dynamical Systems

It is a fundamental fact and a direct consequence of the Theorem 2.1 that there exists a unique of solution for the initial value problem of linear dynamical systems.

Theorem 2.2 *Let A be an $n \times n$ matrix. Then for a given $x_0 \in \mathbb{R}^n$, the initial value problem*

$$\dot{x} = Ax, \ x(t_0) = x_0 \tag{2.2}$$

has a unique solution given by

$$x(t) = e^{At} x_0. \tag{2.3}$$

Proof For the existence, it is enough to show that 2.3 in fact satisfies the Eq. 2.2. This follow directly from $\frac{d}{dt} e^{At} = A e^{At}$. Thus,

$$\dot{x}(t) = \frac{d}{dt} e^{At} x_0 = A e^{At} x_0 = Ax(t).$$

Therefore, $x(t)$ is a solution. To show the uniqueness, lets assume that $x(t)$ is an arbitrary solution of the Eq. 2.2 and lets define $y(t) = e^{-At} x(t)$. From

$$\dot{y}(t) = -Ae^{-At}x(t) + e^{-At}\dot{x}(t)$$
$$= -Ae^{-At}x(t) + e^{-At}Ax(t)$$
$$= 0 \tag{2.4}$$

follows that $y(t) = const.$ and if $t = 0$ then $y(t) = x_0$. These imply simply that

$$x(t) = e^{At}y(t) = e^{At}x_0. \qquad \square$$

Definition 2.9 The set of mappings $e^{At} : \mathbb{R}^n \rightarrow \mathbb{R}^n$ describes the motion of the points $x_0 \in \mathbb{R}^n$ along the trajectories of 2.2 in the phase space and is called the flow of the linear system.

Definition 2.10 If all eigenvalues of the $n \times n$ matrix A have non-zero real parts, then the flow $e^{At} : \mathbb{R}^n \rightarrow \mathbb{R}^n$ is called a hyperbolic flow and 2.2 is called a hyperbolic linear system.

Definition 2.11 A subspace $E \subset \mathbb{R}^n$ is called e^{At}-invariant, if $e^{At}E \subset E$ for all $t \in \mathbb{R}$.

Based on these definitions, we can now introduce the concept of stability for linear dynamical systems.

Definition 2.12 Consider a linear system

$$\dot{x} = Ax, \tag{2.5}$$

with $x \in \mathbb{R}^n$ and A a (real) $n \times n$ matrix with the generalized eigenvectors $v_j = s_j + it_j$ corresponding to an eigenvalue $\lambda_j = a_j + ib_j$. And let

$$B = \{s_1, \ldots, s_k, s_{k+1}, t_{k+1}, \ldots, s_m, t_m\}$$

represent a basis of $\mathbb{R}^n$ with $n = 2m - k$. Then the $\mathbb{R}^n$-subspaces

$$E^s = Span\{s_j, t_j | a_j < 0\}, \tag{2.6}$$
$$E^u = Span\{s_j, t_j | a_j = 0\}, \tag{2.7}$$
$$E^c = Span\{s_j, t_j | a_j > 0\} \tag{2.8}$$

are called the stable, unstable and center subspaces (or manifolds) of the dynamical system.

Definition 2.13 If all eigenvalues of the matrix A have negative (positive) real parts, then the origin is called a sink (source) for the linear system.

Theorem 2.3 *Let A be a real $n \times n$ matrix. Then*

$$\mathbb{R}^n = E^s \oplus E^u \oplus E^c \tag{2.9}$$

where E^s, E^u and E^c are stable, unstable and center subspaces of the linear system 2.5 respectively. Furthermore, the subspaces E^s, E^u and E^c are e^{At}-invariant.

In other words, if a solution of the dynamical system starts in E^s, E^u or E^c at time $t = 0$, then it remains stable, unstable or periodic for all time.

Proof Since $B = \{s_1, \ldots, s_k, s_{k+1}, t_{k+1}, \ldots, s_m, t_m\}$ is a basis of $\mathbb{R}^n$, then from the definitions of E^s, E^u and E^c follows $\mathbb{R}^n = E^s \oplus E^u \oplus E^c$. To show the flow invariance of the subspaces, we will focus on the stable subspace as the proofs of other cases are similar. Now, if $x_0 \in E^s$, then $x_0 = \sum_{j=1}^{n_s} c_j V_j$, where $V_j = s_j$ or t_j and $\{V_j\}_{j=1}^{n_s} \subset B$ is a basis of E^s. Then by linearity of e^{At} follows

$$e^{At} x_0 = \sum_{j=1}^{n_s} c_j e^{At} V_j.$$

Since $e^{At} V_j = \lim_{k \to \infty} (\mathbb{I} + At + \cdots + \frac{A^k t^k}{k})$ for $j = 1, \ldots, n_s$ and V_j is a generalized eigenvector of A, then $A V_j \in E^s$ and $A^k V_j \in E^s$. Thus, for all $t \in \mathbb{R}$, $e^{At} x_0 \in E^s$ and therefore $e^{At} E^s \subset E^s$, i.e. E^s is e^{At}-invariant. $\qquad\square$

In $\mathbb{R}^2$, these subspaces and the stability behavior of linear dynamical systems become simple geometric interpretations as the next theorem shows.

Theorem 2.4 *Considering the dynamical system $\dot{x} = Ax$ (Eq. 2.5), $d = det\ A$ and $c = trace\ A$, then*

- *If $d < 0$, then the system has a saddle in origin.*
- *If $d > 0$ and $c^2 - 4d \geq 0$, then the Eq. 2.5 has a node at the origin. It is stable if $c < 0$ and unstable if $c > 0$.*
- *If $d > 0$, $c^2 - 4d < 0$ and $c \neq 0$, then the Eq. 2.5 has a focus at the origin. It is stable if $c < 0$ and unstable if $c > 0$.*
- *If $d > 0$ and $c = 0$, then the system has a center at the origin.*

Proof The eigenvalues of the matrix A are given by $\lambda = \frac{c \pm \sqrt{c^2 - 4d}}{2}$. Therefore, if $d < 0$, then there exist two real eigenvalues of opposite sign. If $d > 0$ and $c^2 - 4d \geq 0$, then there are two real eigenvalues of the same sign as c. If $d > 0$, $c^2 - 4d < 0$ and $c \neq 0$, then there exist two complex conjugated eigenvalues and finally, if $d > 0$ and $c = 0$, then there are two pure imaginary complex conjugated eigenvalues. $\qquad\square$

The reader is advised to consult further literature on the theory of ordinary differential equations for the graphical representation of these results as saddles, nodes, foci and centers.

2.3 Stability of Non-linear Dynamical Systems

In the non-linear theory, the sign of real parts of the eigenvalues λ_i of the derivative matrix $Df(x_0)$ of the vector field f of the dynamical system $\dot{x} = f(x)$ at any hyperbolic equilibrium point x_0 will determine the stability behavior. If $Re(\lambda_i) < 0$ for $i = 1, \ldots, n$, then a hyperbolic equilibrium point is asymptotically stable and x_0 is a sink. And x_0 is unstable if it is either a source or a saddle. However, the stability of non-hyperbolic equilibria is usually more difficult to determine. Therefore, we discuss briefly a method introduced by Lyapunov to investigate stability behavior of the non-hyperbolic equilibrium points.

We have already introduced the concept of Lyapunov stability by the Definition 2.5. Here, we revamp the previous definition in context of flows of non-linear dynamical system, which is a necessary step for further investigations:

Definition 2.14 Let ϕ_t denote the flow of $\dot{x} = f(x)$ defined for all $t \in \mathbb{R}$. An equilibrium point is called x_0 is called stable if for any $\varepsilon > 0$ there is a $\delta > 0$ such that for all $x \in N_{x_0}(\delta)$ and $t \geq 0$ we have

$$\phi_t(x) \in N_{x_0}(\delta).$$

The equilibrium point is called unstable if it does not fulfill this property. And it is called asymptotically stable if it is stable and there exists a $\delta > 0$ such that for each $x \in N_{x_0}(\delta)$ we have

$$\lim_{t \to \infty} \phi_t(x) = x_0.$$

Definition 2.15 A point $p \in E$ is called an ω-limit point of the trajectory $\phi(\cdot, x)$ of the dynamical system $\dot{x} = f(x)$ if there is a sequence $t_n \to \infty$ such that

$$\lim_{n \to \infty} \phi(t_n, x) = p.$$

Similar, if there exists a sequence $t_n \to -\infty$ with

$$\lim_{n \to \infty} \phi(t_n, x) = q.$$

then q is called an α-limit point.

Definition 2.16 A limit cycle Γ of a two-dimensional system is a cycle of the dynamical system that is the α- or the ω-limit set of some trajectories.

Theorem 2.5 *If x_0 is a sink of the non-linear system and $Re(\lambda_i) < -a < 0$ for all eigenvalues λ_i of $Df(x_0)$, then given $\varepsilon > 0$, there exists a $\delta > 0$ such that for any $x \in N_a(\delta)$ the flow of the system satisfies*

$$|\phi_t - x_0| \leq \varepsilon e^{-at} \ \forall \, t \geq 0.$$

Theorem 2.6 *If x_0 is a stable equilibrium point of $\dot{x} = f(x)$, then no eigenvalue of $Df(x_0)$ has positive real parts.*

It is now clear that the only stable equilibrium point that are not asymptotically stable are of non-hyperbolic type. However, we have still not answered whether a non-hyperbolic equilibrium point is stable, asymptotically stable or unstable. Using the method developed by Lyapunov (below), we will try to address this question.

Definition 2.17 Let E be an open subset of $\mathbb{R}^n$, $f \in C^1(E)$, $L \in C^1(E)$ and ϕ_t the flow of the dynamical system, then the derivate of the function $L(x)$ for $x \in E$ along the solution $\phi_t(x)$ is

$$\dot{L}(x) = \frac{d}{dt} L(\phi_t(x))|_{t=0} = DL(x)f(x). \tag{2.10}$$

Theorem 2.7 *Let $E \subset \mathbb{R}^n$ an open subset containing x_0 and suppose the vector field $f \in C^1(E)$ and $f(x_0) = 0$. Furthermore, let us assume that there exists a Lyapunov function $L : \mathbb{R}^n \to \mathbb{R}$, i.e. a real valued function $L \in C^1(E)$ with $L(x_0) = 0$ and $L(x) > 0$ if $x \neq x_0$. Then the following are true:*

1. *If $\dot{L}(x) \leq 0$, $\forall\, x \in E$, then x_0 is stable.*
2. *If $\dot{L}(x) < 0$, $\forall\, x \in E \setminus \{x_0\}$, then x_0 is asymptotically stable.*
3. *If $\dot{L}(x) > 0$, $\forall\, x \in E \setminus \{x_0\}$, then x_0 is unstable.*

Proof Suppose, without loss of generality, that the equilibrium point is $x_0 = 0$.

(1) Let us choose $\varepsilon > 0$ small enough such that the closer of the ε-neighborhood $\overline{N_0(\varepsilon)} \subset E$ and let m_ε be the minimum of the Lyapunov function $L(x)$ on the compact sphere $S_\varepsilon = \{x \in \mathbb{R}^n \,||x| = \varepsilon\}$. Since $L(x)$ is continuous and $L(0) = 0$, there exists a $\delta > 0$ such that $|x| < \delta$ implying $L(x) < m_\varepsilon$. Furthermore, since $\dot{L}(x) \leq 0$ for $x \in E$, $L(x)$ is decreasing along the trajectories of the dynamical system. Thus, for the flow ϕ_t and for all $x_0 \in N_0(\varepsilon)$ and $t \geq 0$, we have

$$L(\phi_t(x_0)) \leq L(x_0) < m_\varepsilon . \tag{2.11}$$

Now, suppose that for $|x_0| < \delta$ there exists a $t_1 > 0$ with $|\phi_{t_1}(x_0)| = \varepsilon$, i.e. $\phi_{t_1}(x_0) \in S_\varepsilon$. Since m_ε is the minimum of $L(x)$ on S_ε, then $L(\phi_{t_1}(x_0)) \geq m_\varepsilon$, which contradicts the inequality 2.11. Therefore, $|x_0| < \delta$ and $t \geq 0$ implies $|\phi_{t_1}(x_0)| < \varepsilon$, i.e. 0 is a stable equilibrium point.

(2) Now, let $\dot{L}(x) < 0$ for all $x \in E$. Then $L(x)$ is strictly decreasing along the trajectories. If $|x_0| < \delta$, then $\phi_t(x_0) \subset N_0(\varepsilon)$ for all $t \geq 0$. Suppose, $\{t_k\}$ is an arbitrary sequence with $t_k \to \infty$. Since $\overline{N_0(\varepsilon)}$ is compact, there exists a subsequence of $\{\phi_{t_k}(x_0)\}$ converging to a point in $\overline{N_0(\varepsilon)}$. Since $L(x)$ is continuous and strictly decreasing along trajectories, it follows that for any such subsequence $\{t_n\}$ of $\{t_k\}$ the limit is zero and therefore $\phi_{t_k}(x_0) \to 0$ for all $t_k \to \infty$, thus $\phi_t(x_0) \to 0$ for $t \to \infty$, i.e. the equilibrium point 0 is asymptotically stable.

(3) Let M be the maximum of $L(x)$ on $\overline{N_0(\varepsilon)}$. Since $\dot{L}(x) > 0$, $L(x)$ is strictly increasing along the trajectories. Thus, for any $\delta > 0$ and $x_0 \in N_0(\varepsilon) \setminus \{0\}$, we have

$L(\phi_t(x_0)) > L(x_0) > 0$ for all $t > 0$. Since $\dot{L}(x)$ is positive definite, it implies $\inf_{t \geq 0} \dot{L}(\phi_t(x_0)) = m > 0$. Thus, $L(\phi_t(x_0)) - L(x_0) \geq mt$ for all $t \geq 0$ and therefore $L(\phi_t(x_0)) > mt > M$ for sufficiently large t, i.e. $\phi_t(x)$ is outside of the closed set $\overline{N_0(\varepsilon)}$ and 0 is unstable.   □

2.4 Bifurcation Theory

In this section we discuss the role of changes in the vector fields f on the qualitative behavior of the dynamical systems $\dot{x} = f(x)$. For this purpose, we use the concept of structural stability that was first introduced by Andronov and Pontryagin in the last 30s of twentieth century. In general, structural stability refers to the robustness of the qualitative behavior of the system in presence of perturbations of the vector fields. In other words, f is called structurally stable if for any vector field g close to f, the vector fields f and g are topologically equivalent. Structurally stable vector fields on a compact, two-dimensional manifolds are completely characterized (Peixoto's Theorem (1962)), while in higher dimensions ($n \geq 3$) the case is not that clear. Vector fields $f \in C^1(E)$ that are not structurally stable belong to the so-called bifurcation set in $C^1(E)$. The qualitative structure of the solution sets changes as the vector field f passes through a point in the bifurcation set. This is a phenomena that often occurs in biological processes that marks the transition towards biological switches and hysteresis in this book. We will study different types of bifurcations that occur in C^1-systems

$$\dot{x} = f(x, \mu) \qquad (2.12)$$

depending on a parameter $\mu \in \mathbb{R}^n$. In particular, we our attention is directed towards local bifurcations at non-hyperbolic equilibrium points and periodic orbits including bifurcations of periodic orbits from non-hyperbolic equilibrium points.

Definition 2.18 Let E be an open subset of $\mathbb{R}^n$. A vector field f is called structurally stable if there is an $\varepsilon > 0$ (ε-perturbation) such that for all $g \in C^1(E)$ with

$$\| f - g \|_1 < \varepsilon,$$

f and g are topologically equivalent,i.e. there exists a homeomorphism $T : E \circlearrowleft$, which maps the trajectories of $\dot{x} = f(x)$ onto the trajectories of $\dot{x} = g(x)$ and preserves their orientation. In this case we call the dynamical system structurally stable.

If a vector field is not structurally stable, then it is called structurally unstable. Now, let us consider a more general definition of structural stability that become useful in the characterization of structurally stable manifolds.

Definition 2.19 Let f be a C^1-vector field on a compact, n-dimensional differentiable manifold M. Then f is structurally stable on M if there exists an $\varepsilon > 0$ such that for any $g \in C^1(M)$ with $\| f - g \|_1 < \varepsilon$, f and g are topologically equivalent.

Theorem 2.8 *Let $f \in C^1(E)$ with $E \subset \mathbb{R}^n$ an open subset containing a hyperbolic equilibrium point x_0 of $\dot{x} = f(x)$. Then for any $\varepsilon > 0$ there is a $\delta > 0$ such that for all $g \in C^1(E)$ with $\| f - g \|_1 < \delta$, there exists a $y_0 \in N_{x_0}(\varepsilon)$ such that y_0 is a hyperbolic equilibrium point of $\dot{x} = g(x)$. Furthermore, f and g have the same number of eigenvalues with positive and negative real parts.*

A direct consequence of this theorem is for linear systems $\dot{x} = Ax$ where A has no eigenvalues with zero real parts. Such a linear dynamical system is always structurally stable. This theorem can easily be formulated for periodic orbits.

Theorem 2.9 *Let $f \in C^1(E)$ with $E \subset \mathbb{R}^n$ an open subset containing a hyperbolic periodic orbit Π_f of $\dot{x} = f(x)$. Then for any $\varepsilon > 0$ there is a $\delta > 0$ such that for all $g \in C^1(E)$ with $\| f - g \|_1 < \delta$, there exists a $\Pi_g \in N_{\Pi_f}(\varepsilon)$ such that Π_g is a hyperbolic periodic orbit of $\dot{x} = g(x)$. Furthermore, the stable and unstable manifolds of Π_f and Π_g have equal dimensions, respectively.*

Definition 2.20 A point $x \in M$ is a non-wandering point of the flow ϕ_t if for any neighborhood N_x and any $T > 0$ there is a $t > T$ such that

$$\phi_t(N_x) \cap N_x \neq \emptyset.$$

This definition is essential for the Peixoto theorem (below). Equilibrium points and points on periodic orbits are examples of non-wandering points. However, the converse is in general not true.

Theorem 2.10 **(Peixoto)** *Let f be a C^1 vector field on a compact, two-dimensional differentiable manifold M. Then f is structurally stable on M if and only if*

- *the number of equilibrium points and cycles is finite and each is hyperbolic;*
- *there are no trajectories connecting saddle points;*
- *the set of non-wandering points only contains equilibrium points and limit cycles.*

Furthermore, if M is orientable, then the set of structurally stable vector fields in $C^1(M)$ is an open, dense subset of $C^1(M)$.

Corllary 2.1 *The set of structurally stable vector fields in on the two-dimensional sphere or any connected sum of tori is an open, dense subset of C^1 on the same manifold.*

Polynomial vector fields are of great importance for mathematical modeling of biological processes, as such vector fields are often obtained by nonlinear regression analysis of experimental data. To extend the Peixoto's theorem on polynomial vector fields (Kotus et al. 1982), we need to introduce a few further concepts, particularly it is necessary to define a saddle connection.

Definition 2.21 A saddle at infinity of a planar real vector field f is a pairing (Σ_p^+, Σ_q^-) of half-trajectories of f, each escaping at infinity such that there exist sequences $p_n \to p$ and $t_n \to \infty$ with $\phi(t_n, p_n) \to q$ in $\mathbb{R}^2$. Σ_p^+ and Σ_q^- are called the stable and unstable separatrix of the saddle at infinity respectively. Based on these definitions, a saddle connection is a trajectory Σ of f defined by $\Sigma = \Sigma^+ \cup \Sigma^-$.

Definition 2.22 A C^1 vector field f is structurally stable on $\mathbb{R}^2$ under strong C^1-perturbations if it is topologically equivalent to any C^1 vector field g satisfying

$$|f(x) - g(x)| + \| Df(x) - Dg(x) \| < \varepsilon(x),$$

for a continuous, strictly positive function $\varepsilon(x)$ on $\mathbb{R}^2$.

Theorem 2.11 *A polynomial vector field is structurally stable on $\mathbb{R}^2$ under strong C^1-perturbations if and only if*

- *all equilibrium points and cycles are hyperbolic;*
- *there are no saddle connections.*

The transition from planar systems towards higher dimensional vector fields is not trivial. However, the Morse-Smale systems on compact, n-dimensional differentiable manifolds are known to be structurally stable although the converse is in general incorrect for dimensions $n \geq 3$.

Definition 2.23 A Morse-Smale system is defined by

- the number of equilibrium points and cycles is finite;
- if $p \in E^s \cap E^u$, then $T_p E^s \oplus T_p E^u = \mathbb{R}^n$, with $T_p E^s$ and $T_p E^u$ the tanget spaces of E^s and E^u respectively at p;
- the set of non-wandering points only contains equilibrium points and limit cycles.

Let us recall that the qualitative behavior of the solution set of $\dot{x} = f(x, \mu)$ depending on the parameter $\mu \in \mathbb{R}$ (or $\mu \in \mathbb{R}^n$), changes with the variations of the parameter μ through a so-called bifurcation value (defined below).

Definition 2.24 A bifurcation value μ_0 of the parameter μ in $\dot{x} = f(x, \mu)$, is a specific value for which the C^1 vector field $f(x, \mu_0)$ is not structurally stable.

Here, we confine on an exemplary discussion of different types of bifurcations that occur in one-dimensional systems with a view towards bistability and hysteresis and finish our excursion on the stability and bifurcation theory. In the following, f is considered as a C^1 vector field on $E \times J$, with $E \subset \mathbb{R}^n$ an open subset and $J \subset \mathbb{R}$ an interval.

Saddle-node bifurcations: Let us consider the equation $\dot{x} = \mu - x^2$. For $\mu > 0$, the system possesses two equilibrium points $x = \pm\sqrt{\mu}$ with $Df(\pm\sqrt{\mu}, \mu) = \mp 2\sqrt{\mu}$. Thus, the equilibrium point at $x = \sqrt{\mu}$ is stable and $x = -\sqrt{\mu}$ is unstable. For $\mu = 0$, there exists only one equilibrium point at $x = 0$ and it is a non-hyperbolic one since $Df(0, 0) = 0$. The vector field $f(x) = -x^2$ is

Fig. 2.1 The bifurcation diagram of the one-dimensional saddle-node bifurcations. The *solid* and *dashed curves* represent the stable and unstable equilibrium points respectively. $\mu = 0$ is a bifurcation value

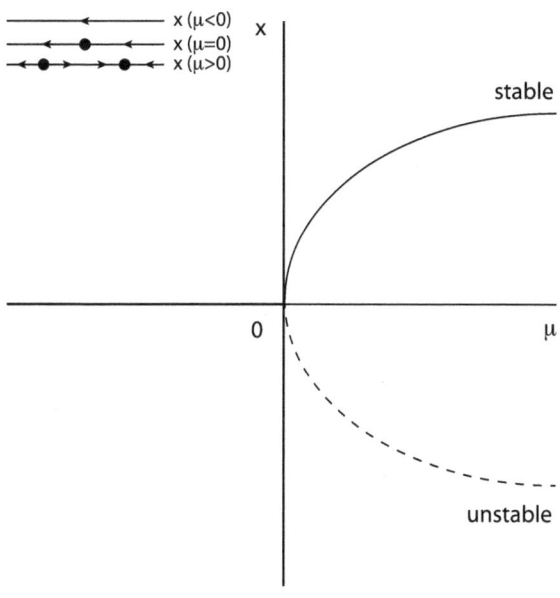

Fig. 2.2 The bifurcation diagram of the one-dimensional transcritical bifurcations. The *solid* and *dashed curves* represent the stable and unstable equilibrium points respectively. $\mu = 0$ is a bifurcation value

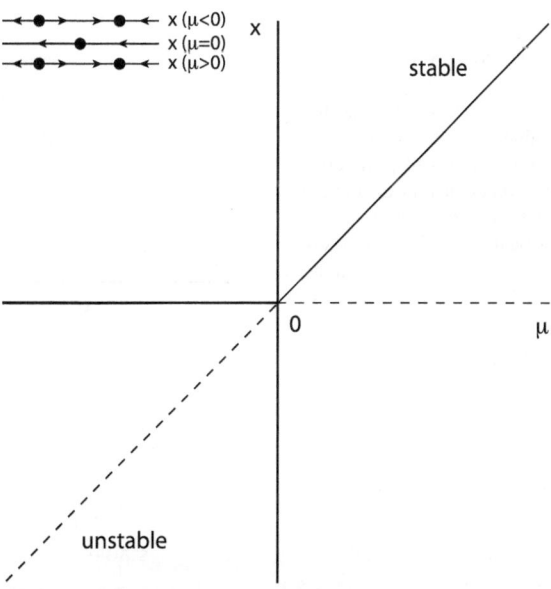

structurally unstable and $\mu = 0$ is a bifurcation value. There are no equilibrium points for $\mu < 0$. All of the relevant information concerning bifurcation behavior with respect to parameter μ is represented in the bifurcation diagram in Fig. 2.1.

Fig. 2.3 The bifurcation diagram of the one-dimensional pitchfork bifurcations. *The solid* and *dashed curves* represent the stable and unstable equilibrium points respectively. $\mu = 0$ is a bifurcation value

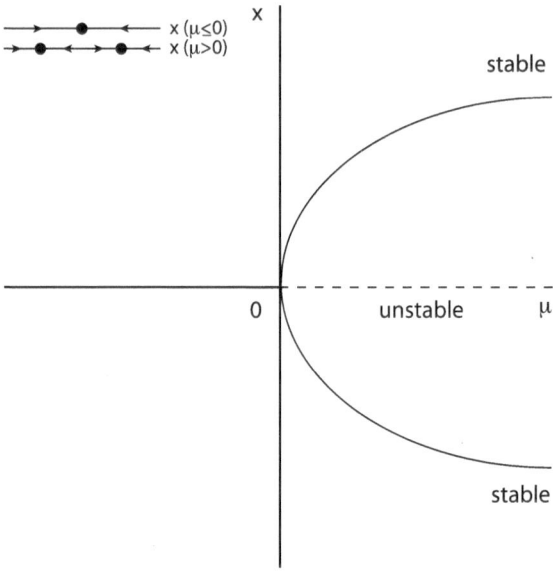

Fig. 2.4 The bifurcation diagram of the one-dimensional hysteresis-like bifurcations. *The solid* and *dashed curves* represent the stable and unstable equilibrium points respectively. The *black arrows* denote the motion starting below the critical value $-\mu_c$ and show the switch at μ_c to the higher equilibrium states, whereas the *grey arrows* show the reverse motion

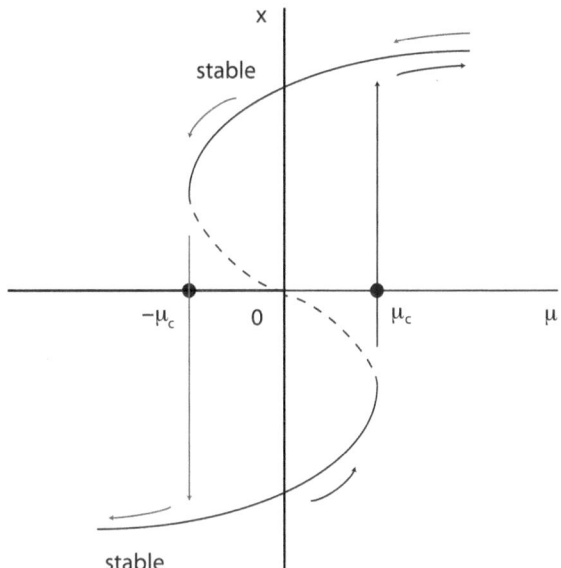

The stable and unstable manifolds are given by $E^s(\sqrt{\mu}) = (-\sqrt{\mu}, \infty)$ and $E^u = (-\infty, \sqrt{\mu})$. For $\mu = 0$, the center manifold is given by $E^c = (-\infty, \infty)$.
Transcritical bifurcations: Consider the dynamical system $\dot{x} = \mu x - x^2$. The equilibrium points are at $x = 0$ and $x = \mu$. Similar to saddle-node bifurcations for $\mu = 0$, there is only one equilibrium point at $x = 0$, which is non-hyperbolic

$(Df(0, 0) = 0)$. $f(x) = -x^2$ is structurally unstable and therefore, $\mu = 0$ is a bifurcation value with the center manifold $E^c = (-\infty, \infty)$. Figure 2.2 demonstrates the exchange of stability behavior at $\mu = 0$.

Pitchfork bifurcations: Let us consider $\dot{x} = \mu x - x^3$. This equation reveals the main mechanism for switches between mono- and bistability in non-linear systems. For $\mu > 0$, the equilibrium points are at $x = 0$ and $x = \pm\sqrt{\mu}$. For $\mu \leq 0$, $x = 0$ is the sole equilibrium point, non-hyperbolic at $x = 0$ $(Df(0, 0) = 0)$. $f(x) = -x^3$ is structurally unstable and $\mu = 0$ a bifurcation value (Fig. 2.3). If $\mu < 0$, then the stable manifold is given by $E^s = (-\infty, \infty)$. For $\mu = 0$ however, the stable manifold is the empty set and the center manifold $E^c = (-\infty, \infty)$.

Hysteresis-like bifurcations: Lets consider the non-linear equation $\dot{x} = \mu + x - x^3$. The bifurcation diagram in Fig. 2.4 of this dynamical system shows the bistability, switch-behavior and hysteresis non-linearity simultaneously in a manner that is similar to many biological applications.

If the system starts with a parameter value below the critical value $-\mu_c$, then the system follows the stable path, however when it reaches the mirrored critical value μ_c, the branch does not continue and the system jumps to the higher branch. The backwards procedure is similar but converse.

Such an on-off (binary) switch mechanism is the simplest form of hysteresis and provides the foundations of the concept of hysteron as the elementary hysteresis module.

References

Kotus J, Kyrch M, Nitecki Z (1982) Global structural stability of flows on open surfaces. Mem Am Math Soc 37:1–108

Peixoto M (1962) Structural stability on two-dimensional manifolds. Topology 1:101–120

Perko L (2008) Differential equations and dynamical systems. Springer, Heidelberg

Verhulst F (2008) Nonlinear differential equations and dynamical systems. Springer, Heidelberg

Chapter 3
Preisach Models

3.1 Definition and Main Properties of Preisach Hysteresis Operators

This chapter intends to familiarize the readers with the Preisach model of hysteresis. Since its publication in the 1930s of the last century (Preisach 1935), the model has been further developed and improved and many valuable facts have been accumulated (Everett and Whitton 1952; Everett 1954, 1955; Enderby 1956; Biorci and Pescetti 1958, 1959, 1966; Brown 1962; Bate 1962; Woodward and Della Torre 1960; Della Torre 1965; Damlanian and Visintin 1983; Visintin 1984; Barker et al. 1985; Brokate and Visintin 1989; Krasnosel'skii and Pokrovskii 1989). As a result of ages of interdisciplinary work on hysteresis phenomena, a stringent mathematical and universal definition of hysteresis is lacking. Therefore, we begin with a general definition of hysteresis in order to avoid any confusion and ambiguity.

A non-linear multi-branch operator is called a hysteresis operator, if it transforms the extreme values of the input functions into branch transitions. In the following, we will only consider rate-independent hysteresis operators, for which the branches of the non-linearities are solely determined by the memory of input extrema and not on the speed of input variations between extreme points.

Similar to the spectral decomposition of self-adjoint operators in functional analysis, hysteresis operators in Preisach model can be represented as a weighted superposition of elementary relay operators with local memories called hysterons. It should be noticed for any hysteresis operator that it is be no means a priori trivial how to select admissible input spaces. In this book, we will begin with the study of hysteresis outputs corresponding to continuous input functions and then extend the discussion onto different biologically relevant admissible function spaces.

H. R. Noori, *Hysteresis Phenomena in Biology*, SpringerBriefs in Mathematical Methods, DOI: 10.1007/978-3-642-38218-5_3, © The Author(s) 2014

Definition 3.1 Let $x \in C^0$ be an arbitrary continuous real-valued function and $\eta_0 \in \{-1, 1\}$ a normalized initial state (sometimes $\{0, 1\}$). Then $h_{\alpha,\beta}[t_0, \eta_0]$ for $t \geq t_0$ is called the hysteron operator (or sometimes non-ideal relay operator) if

$$
y(t) = h_{\alpha,\beta}[t_0, \eta_0]x(t) = \begin{cases} \eta_0, & \text{if } \alpha < x(\tau) < \beta \ , \ \forall \tau \in [t_0, t] \\ 1, & \text{if } \exists t_1 \in [t_0, t] \ : \ x(t_1) \geq \beta \ , \ x(\tau) > \alpha \ \forall \tau \in [t_1, t] \\ -1, & \text{if } \exists t_1 \in [t_0, t] \ : \ x(t_1) \leq \alpha \ , \ x(\tau) < \beta \ \forall \tau \in [t_1, t] \end{cases}
$$
$$(3.1)$$

where α and β are the threshold values with $\alpha \geq \beta$.

Let us investigate some properties of the hysterons before we introduce the general hysteresis operators of Preisach type.

Theorem 3.1 *If $h_{\alpha,\beta}[t_0, \eta_0]$ is an operator mapping C^0 into L_q for $1 \leq q < \infty$, then it is locally compact.*

Since oscillations occur in many biological phenomena, it is of particular interest to study the analytic behavior of hysteresis operators, particularly hysterons, with periodic inputs.

Let us first define for any continuous input function $x(t)$ two continuous functions by

$$
h_{-1}x(t) = \lim_{\tau \to -\infty} h_{\alpha,\beta}[\tau, -1]x(t)
$$

$$
h_1 x(t) = \lim_{\tau \to -\infty} h_{\alpha,\beta}[\tau, 1]x(t)
$$

Both limits exists. Moreover, for any $\tilde{t}$ there is a $\tilde{\tau}$ such that for $\tau \leq \tilde{\tau}$,

$$
h_{\alpha,\beta}[\tau, -1]x(t) \equiv h_{\alpha,\beta}[\tau, 1]x(t) \tag{3.2}
$$

at $t \geq \tilde{t}$. If the input function $x(t)$ is periodic with the period T, then both functions are periodic with the same period. Furthermore, for the T-periodic input $x(t)$, the corresponding output $y(t) = h_{\alpha,\beta}[t_0, \eta_0]x(t)$ is periodic i.e. $y(t + T) = y(t)$ at $t \geq T + t_0$.

We will now briefly discuss the closer and convexification of the hysterons as it may be useful for specific applications in computational biology. In the following, A will denote the set of all admissible states of a hysteresis operator.

For any continuous input function $x(t)$ ($t \geq t_0$), we define two corresponding sets $\bar{h}_{\alpha,\beta}[t_0, \eta_0]x(t)$ (η as above either -1 or 1) of binary output functions $y(t)$ such that

- if $x(t) > \beta$ at $t \in [t_1, t_2] \subset [t_0, \infty)$, then $y(t)$ is non-decreasing on $[t_1, t_2]$;
- if $x(t) < \alpha$ at $t \in [t_1, t_2] \subset [t_0, \infty)$, then $y(t)$ is non-decreasing on $[t_1, t_2]$;
- $\{x(t), y(t)\} \in A$ at $(t \geq t_0)$ and $y(t_0) = h_{\alpha,\beta}[t_0, \eta_0]x(t_0)$.

Proposition 3.1 *If $h_{\alpha,\beta}[t_0, \eta_0]$ are operators from C^0 to L_q ($1 \leq q < \infty$), then the operators $\bar{h}_{\alpha,\beta}[t_0, \eta_0]$ represent their closers.*

It should be remarked that the closers $\bar{h}_{\alpha,\beta}[t_0, \eta_0]$ differ in their qualitative behavior from the corresponding hysterons.

Definition 3.2 Now, let us consider $h_{\alpha,\beta}[t_0, \eta_0] : C^0 \to L_q$ on an interval spanned by t_0 and t_1. A convexification of the hysteron is then defined by

$$\tilde{h}_{\alpha,\beta}[t_0, \eta_0]x(t) = \bigcap_{\varepsilon > 0} \overline{co}\{y(t)|h_{\alpha,\beta}[t_0, \eta_0]u(t), \| u(t) - x(t) \|_{t_0,t_1} < \varepsilon\}. \quad (3.3)$$

Theorem 3.2 *Each convexification set of the hysteron operator $h_{\alpha,\beta}[t_0, \eta_0]$ is comprised by the functions $y(t)$ satisfying:*

- *if $x(t) > \beta$ at $t \in [\tau_1, \tau_2] \subset [t_0, \infty)$, then $y(t)$ is non-decreasing on $[\tau_1, \tau_2]$;*
- *if $x(t) < \alpha$ at $t \in [\tau_1, \tau_2] \subset [t_0, \infty)$, then $y(t)$ is non-decreasing on $[\tau_1, \tau_2]$;*
- *$\{x(t), y(t)\} \in \tilde{A}$ at ($t \geq t_0$) and $y(t_0) = h_{\alpha,\beta}[t_0, \eta_0]x(t_0)$.*

As mentioned above, the Preisach model of hysteresis can be interpreted as a discontinuous or continuous parallel connection of the hysteron (or relay) operators:

Definition 3.3 Let $h^i = h_{\alpha_i,\beta_i}$ with $1 \leq i \leq N$ for some $N \in \mathbb{N}$ and consider positive weight functions $\mu : \{1, \ldots, N\} \to \mathbb{R}^+$ with $\mu_i = \mu(i)$. Then the parallel connection of hysterons is called the Weiss-Preisach or discontinuous operator $\hat{H}[t_0, \eta_0]$ for $t \geq t_0$, which is given by

$$y(t) = \hat{H}[t_0, \eta_0]x(t) = \sum_{i=1}^{N} \mu_i h^i[t_0, \eta_0(i)]x(t). \quad (3.4)$$

Now, consider a family $\mathscr{R}$ of hysterons $h^\omega = h_{\alpha_\omega,\beta_\omega}$ with threshold values α_ω and β_ω indexed by a measurable space (Ω, μ). We call $\mathscr{R}$ a bundle of hysterons and any measurable function $\eta_0 : \Omega \to \{-1, 1\}$ the initial state of the bundle $\mathscr{R}$.

Definition 3.4 For any initial state $\eta_0(\omega)$ and any continuous input $x(t)$ ($t \geq t_0$), the Preisach hysteresis operator $H[t_0, \eta_0]$ is defined by

$$y(t) = H[t_0, \eta_0]x(t) = \int_\Omega h^\omega[t_0, \eta_0(\omega)]x(t)d\mu. \quad (3.5)$$

Proposition 3.2 *If the measure μ has a finite support $\{\omega_j\}_{j=1,\ldots,n}$ then $\hat{H} \equiv H$.*

As discussed above, every hysteron is monocyclic, hence the output state of functions $y(t)$ corresponding to a T-periodic input $x(t)$ (for $t \geq t_0$) is T-periodic for $t \geq T + t_0$. Consequently, also the Weiss-Preisach (and Preisach) operators and bundles $\mathscr{R}$ of hysterons are monocyclic. The reader should notice that most of the following results that are formulated for discontinuous operators can be extended to the continuous Preisach operators.

Definition 3.5 The closed curve

$$\mathcal{L} = \{(x, y)|x = x(t), \ y(t) = h[t_0, \eta_0]x(t)\} \tag{3.6}$$

for $t_0 + T \leq t \leq t_0 + 2T$ is called a hysteresis loop.

Proposition 3.3 *If the hysteresis loop $\mathcal{L}$ of hysterons for a fixed period T is completely determined without the initial states then it has a non-zero width, else it is degenerated and can be represented by a curve.*

Definition 3.6 For Weiss-Preisach operators, the hysteresis loop is uniquely defined by

$$\mathcal{L} = \{(x, y)|x = x(t), \ y(t) = \hat{H}[t_0, \eta_0]x(t)\} \tag{3.7}$$

for $t_0 + T \leq t \leq t_0 + 2T$, if all hysterons are uniquely defined by the input functions $x(t)$.

Assume $x(t)$ is a T-periodic input function with only one local maximum and one local minimum on any interval of the length of the period T. Then all curves $\mathcal{L}$ separate a part of the plane, which is also called hysteresis loop.

Theorem 3.3 *Suppose there are two hysteresis loops corresponding to a T-periodic input $x(t)$ for $t \geq t_0$ for the Weiss-Preisach operator. Then the areas of those loops are equal and for any x^*, the intersection of the line $x = x^*$ with both loops are segments of the same length.*

Definition 3.7 The Weiss-Preisach operator is said to have positive spin, if every hysteresis loop corresponding to the input function $x(t)$ possessing only one local maximum and minimum within one period T can be circulated by points $\{(x(t), y(t))\}$ for $t_0 + T \leq t \leq t_0 + 2T$ in counter-clockwise direction. Analogous, negative spin refers to the circulation in clockwise direction.

Definition 3.8 The functional $S[x(t); T]$ is called the oriented are of the hysteresis loop if

$$S[x(t); T] = \int_{t_0+T}^{t_0+2T} h[t_0, \eta_0]x(t)dx(t). \tag{3.8}$$

For Weiss-Preisach operators this definition can be extended to

$$\hat{S}[x(t); T] = \sum_{i=1}^{N} S_i[x(t); T]. \tag{3.9}$$

We can further define the oriented area of the hysteresis loops for any continuous periodic input $x(t)$ by $S[x(t); T] = \lim_{n \to} S[x_n(t), T]$ for continuously differentiable T-periodic functions $x_n(t)$, which uniformly converge to $x(t)$. However, the definition cannot be extended to any smooth almost periodic function as the hysteresis loops do not exist for those functions.

Let us denote by $P[T]$ the operator transforming any T-periodic input $x(t)$ for $t \geq t_0$ into the set $P[T]x(t)$ of all T-periodic outputs ($t \geq T + t_0$) of the Preisach operators.

Theorem 3.4 *If the Weiss-Preisach operator exists, then the continuous single-valued operator mapping any T-periodic input onto the corresponding T-periodic output is a selector of the operator $P[T]$.*

so far, we have been concerned with parallel connections of hysterons. Let us now consider the sequential connection of a finite set of hysterons. In general, the sequential connection of N hysterons h^i induces an operator W with scalar input function $x(t)$ and scalar ouput $y(t)$ called sequential connection, which intrinsically is structured as a recursively defined vector-valued operator Q with values in $\mathbb{R}^N$ that we call a cascade of hysterons. For W and Q the states are given by the vectors $\{x, v_1, \ldots, v_N\} \in \mathbb{R}^{N+1}$ such that $\{x, v_1\} \in A(h^1)$, $\{v_1, v_2\} \in A(h^2)$, ..., $\{v_{N-1}, v_N\} \in A(h^N)$. The hysteresis operators W and Q are then defined recursively by

$$v_1(t) = h^1[t_0, \eta_0(1)]x(t) \qquad (3.10)$$

$$v_2(t) = h^2[t_0, \eta_0(2)]v_1(t) \qquad (3.11)$$

$$\vdots$$

$$y(t) = v_N(t) = h^N[t_0, \eta_0(N)]v_{N-1}(t). \qquad (3.12)$$

Proposition 3.4 *If $x(t)$ is a T-periodic hysteron, then $y(t)$ is T-periodic for $t \geq NT + t_0$ with the hysteresis loop*

$$\{(x, y)|x = x(t), \; y = y(t)\} \qquad (3.13)$$

on $NT + t_0 \leq t \leq (N+1)T + t_0$.

In contrast to the parallel connection of hysterons, the are of hysteresis loops for sequential connections not only depends on the T-periodic input functions but also on the initial states.

In the following, we will introduce a formal characterization of the Weiss-Preisach hysteresis operators under small perturbations of input functions, which is of great importance for mathematical modeling of natural phenomena. The presence of small stochastic perturbations induces a slow and gradual annulation of the stored information in Preisach models, which should be taken into account for modeling and analysis of processes with intrinsic noise. Let us assume that there are no dependencies between a set of perturbations $\{\varepsilon_i(t)\}_{i=1,\ldots,N}$ and the input function $x(t)$, which correspond to the hysterons $h^1, \ldots, h^N$. Then the output function can be characterized by

$$y(t) = \sum_{i=1}^{N} \mu_i h^i[t_0, \eta_0(i)](x(t) + \varepsilon_i(t)). \qquad (3.14)$$

3.2 Geometric Interpretation and Numerical Analysis of Preisach Operators

The geometric interpretation of the Preisach model is based on the fact that there is a unique correspondence between the hysterons $h_{\alpha,\beta}$ and the points (α, β) in the half-plane $\alpha \geq \beta$ (Fig. 3.1).

In light of this representation, the Eq. 3.5 describing the Preisach operator H can easily be reformulated to an equivalent geometric form using the so-called density function value or Preisach density function $\mu(\alpha, \beta)$, which should be determined by use of some experimental data:

$$y(t) = \int\int_{\alpha \geq \beta} \mu(\alpha, \beta) h_{\alpha,\beta}[x_0, \eta_0] x(t) d\alpha d\beta. \qquad (3.15)$$

Now, let us consider the the triangle T in the half-plane $\alpha \geq \beta$ with boundaries defined by the up and down saturation points x_{max} and x_{min} of the input function.

Fig. 3.1 The geometric representation of the Preisach model in the (α, β)-plane. Each point of the half-plane $\alpha \geq \beta$ can be uniquely identified with an operator $h_{\alpha,\beta}$ whose up and down switching values are represented by the α and β coordinates of a specific point. S^+ and S^- refer to those operators with an output equal to 1 or -1 respectively for given instant of time, bounded by the up and down saturation points x_{max} and x_{min}

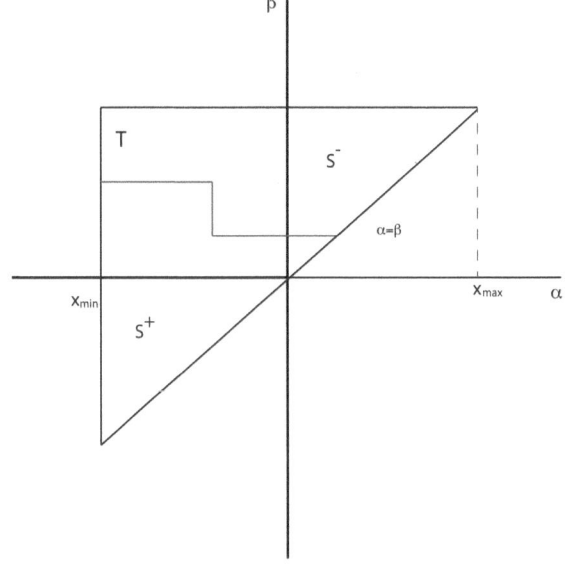

Furthermore, let us without loss of generality assume that *Supp* $\mu(\alpha, \beta) \subseteq T$, i.e. the Preisach density function μ is zero outside of T. Then for any instant of time t, we can split T into two subsets $S^+(t)$ and $S^-(t)$ consisting of points $(\alpha, \beta) \in T$ for which the corresponding hysteron satisfies $y(t) = h_{\alpha,\beta}[x_0, \eta_0]x(t) = 1$ or $y(t) = h_{\alpha,\beta}[x_0, \eta_0]x(t) = -1$, respectively (Fig. 3.1). Based on this decomposition, the Eq. 3.15 can be simplified to

$$y(t) = \int\int_{S^+(t)} \mu(\alpha, \beta)d\alpha d\beta - \int\int_{S^-(t)} \mu(\alpha, \beta)d\alpha d\beta. \qquad (3.16)$$

This description translates the analysis of the output of the Preisach models to the determination of the Preisach density functions $\mu(\alpha, \beta)$ (Identification problem). This simple mathematical form was extended into a numerical form by Mayergoyz (1986, 2003) to implement the Preisach model for experimental applications. In this method, the calculation of the output functions is reduced to a summation of terms that depend on the history of local minima and maxima of the input function.

We will use the set of first-order transition curves for this purpose, which are defined by the following procedure. In order to bring the hysteresis non-linearity to the state of negative saturation, we first decrease the input function below a value β_0. Subsequently, the input is monotonically increased up to some value α' (Fig. 3.2a). We call this branch the limiting ascending branch. We denote with $y_{\alpha'}$ the output value on this branch, which corresponds to the input $x = \alpha'$. To form a first-order transition curve, this monotonic increase of the input is followed by a monotonic decrease. If the output value on the transition curve is attached to the limiting ascending branch at $y_{\alpha'}$ and correspond to the input value $x = \beta'$ (Fig. 3.2b), then we call it $y_{\alpha',\beta'}$.

Based on this procedure, we define the function

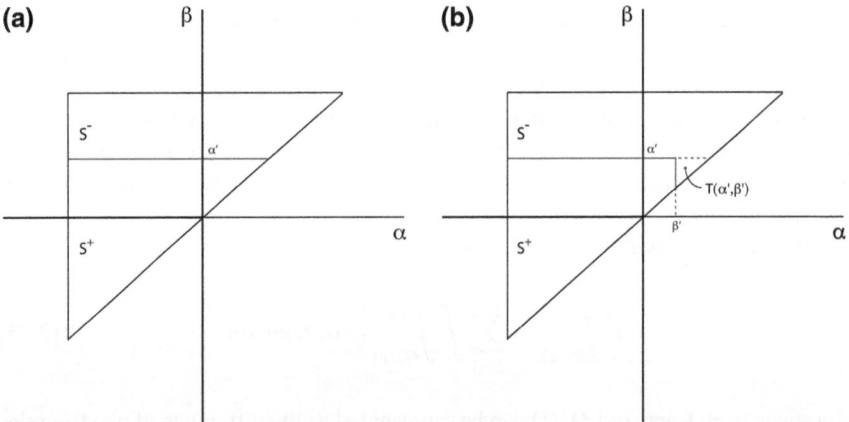

Fig. 3.2 (α, β) plane due to the Mayergoyz procedure for the generation of first-order transition curves

$$F(\alpha', \beta') = \frac{1}{2}(y_{\alpha'} - y_{\alpha', \beta'}), \tag{3.17}$$

which is equal to one-half of the input increments along the first-order transition curves.

As Fig. 3.2b shows, the result of the monotonic input decrease from α' to β' is that visually the triangle $T(\alpha', \beta')$ is added to S^- and subtracted from S^+. Therefore, the Preisach model (3.17) will match the output increments along the first-order transition curves if the Preisach function satisfies

$$y_{\alpha', \beta'} - y_{\alpha'} = -2 \int \int_{T(\alpha', \beta')} \mu(\alpha, \beta) d\alpha d\beta. \tag{3.18}$$

It implies then

$$F(\alpha', \beta') = \int \int_{T(\alpha', \beta')} \mu(\alpha, \beta) d\alpha d\beta, \tag{3.19}$$

$$F(\alpha', \beta') = \int_{\beta'}^{\alpha'} (\int_{\beta}^{\alpha'} \mu(\alpha, \beta) d\alpha) d\beta. \tag{3.20}$$

If we first differentiate this equation with respect to β' and then with respect to α', then

$$\mu(\alpha', \beta') = -\frac{\partial^2 F(\alpha', \beta')}{\partial \alpha' \partial \beta'}, \tag{3.21}$$

$$= \frac{1}{2} \frac{\partial^2 y_{\alpha', \beta'}}{\partial \alpha' \partial \beta'}. \tag{3.22}$$

This formula allows a simple geometric interpretation of the density functions as the first β'-derivative of $y_{\alpha', \beta'}$ is the tangents function of the angle between the axis x and the tangent to the first-order transition curve $y_{\alpha', \beta'}$ at $x = \beta'$.

For the numerical implementation of the above theory, let us consider the positive sets S^+ which can be subdivided into trapezoidal regions $Q_k(t)$ (Fig. 3.3a, b) generated by two pairings (M_k, m_{k-1}) and (M_k, m_k) of maxima M_k and minima m_k of the input history.

Consequently, we have

$$\int \int_{S^+(t)} = \sum_{k=1}^{n(t)} \int \int_{Q_k(t)} \mu(\alpha, \beta) d\alpha d\beta. \tag{3.23}$$

Moreover, each trapezoid $Q_k(t)$ can be represented as the difference of two triangles $T(M_k, m_{k-1})$ and $T(M_k, m_k)$. Thus

$$\int\int_{Q_k(t)} \mu(\alpha, \beta)d\alpha d\beta = \int\int_{T(M_k,m_{k-1})} \mu(\alpha, \beta)d\alpha d\beta - \int\int_{T(M_k,m_k)} \mu(\alpha, \beta)d\alpha d\beta. \tag{3.24}$$

The Eq. 3.20 implies then

$$F(M_k, m_k) = \int\int_{T(M_k,m_k)} \mu(\alpha, \beta)d\alpha d\beta, \tag{3.25}$$

$$F(M_k, m_{k-1}) = \int\int_{T(M_k,m_{k-1})} \mu(\alpha, \beta)d\alpha d\beta. \tag{3.26}$$

Hence,

$$\int\int_{Q_k(t)} \mu(\alpha, \beta)d\alpha d\beta = F(M_k, m_{k-1}) - F(M_k, m_k). \tag{3.27}$$

Therefore, for an decreasing input, as shown in Fig. 3.3b, the final link of boundary interface line is the vertical line $m_n = x(t)$ and output can be calculated as

$$y(t) = \sum_{k=1}^{n(t)-1} (F(M_k, m_{k-1}) - F(M_k, m_k)) + F(M_n, m_{n-1}) - F(M_n, x(t)). \tag{3.28}$$

Consequently, for an decreasing input, as shown in Fig. 3.3a, the final link of boundary interface line is the vertical line $M_n = x(t)$ and output can be calculated as

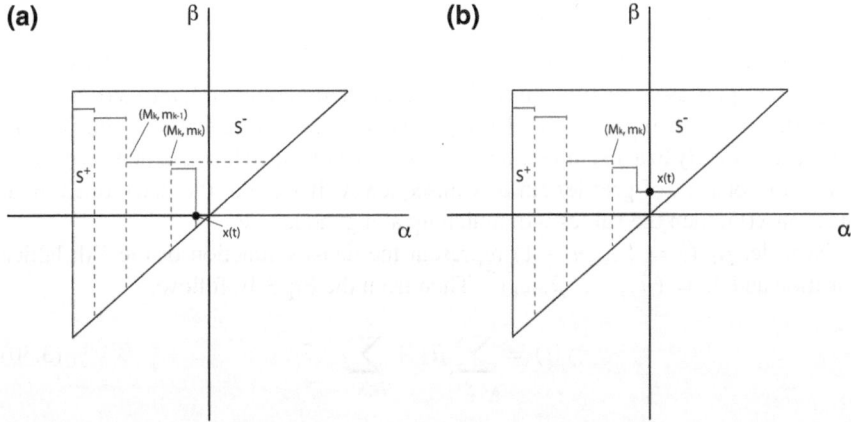

Fig. 3.3 Numerical implementation of the Preisach model for increasing (**a**) and decreasing inputs (**b**)

$$y(t) = \sum_{k=1}^{n(t)-1} (F(M_k, m_{k-1}) - F(M_k, m_k)) + F(x(t), m_{n-1}). \qquad (3.29)$$

These results clearly demonstrate that the output of numerical Preisach model can be calculated from current input value, as well as minima and maxima history terms. The function $F(\alpha, \beta)$ in points (α, β) can be calculated by the application of appropriate interpolation algorithms on the set of experimental data of hysteresis on the first-order reversal curves (ascending and descending). Depending on the application and the properties of the datasets, there are various interpolation algorithms, which can be used for this process such as cubic spline, nearest neighbor, linear interpolation, and Newton method.

3.3 Shirley-Venkataraman Approximation of Preisach Density Functions

The above mentioned Mayergoyz' concept relies on continuous and perturbation-free inputs to exactly predict the density function $\mu(\alpha, \beta)$. Since experimental measurement of biological processes is usually accompanied with some noise, there is a need for an alternative approach. In this section, we present an approximation technique for numerical density function as proposed by Shirley and Venkataraman (2003). which by discretization of the Preisach model, casts the identification problem as a constrained minimization problem

$$\min_{M \geq 0} \frac{1}{2} \parallel AM - Y \parallel^2,$$

where A and Y are computed using the input and output signals, respectively.

For this purpose, we first discretize the input function $x(t)$ into $x(t_i)$ by the discretization of the time domain $[0, T]$ into $t = t_0, \ldots, t_n$ with $t_0 = 0$ and $t_n = T$. This leads directly to a discretization of the $\alpha \geq \beta$ half-plane into a lattice $L = \cup_n L_n$ with partitions L_n of grid length d. Consequently, if we plot the discretized input function $x(t)$, then the values will match up along the lattice.

Now, let $\bar{\mu}_i$ ($i = 1 \ldots n - 1$) represent the density function of the i'th lattice partition and $M = (\bar{\mu}_1, \ldots, \bar{\mu}_{n-1})^T$. Then from the Eq. 3.16 follows

$$y(t) = \sum_{k \in S^+} \bar{\mu}_k + \sum_{l \in S^-} -\bar{\mu}_l. \qquad (3.30)$$

The introduced discretization appears to represent a valid approximation, as the discretized density function $\bar{\mu}$ converges to the density function μ if $n \to \infty$ for L_n.

Proposition 3.5 *Let L_n represent the lattice partitions of the discretization of $\alpha \geq \beta$ half-plane and $y(t)$ the output function of the Preisach operator. Then*

$$\sum_{k \in S^+} \bar{\mu}_k + \sum_{l \in S^-} -\bar{\mu}_l \rightarrow y(t) \tag{3.31}$$

for max $\{\max\{|a - b| : \forall\, a, b \in L_n\}|\ \forall\, L_n\}$.

Now, let us define the vector A as $A = (\delta_1 \cdots \delta_{n-1})$ such that $\delta_i = 1$ if $\bar{\mu}_i \in S^+$ and $\delta_i = -1$ if $\bar{\mu}_i \in S^-$. Then, for a fixed instant of time t_1 we can write the Eq. 3.30 as $y(t_1) = A_1 M = \sum_{i=1}^{n-1} \delta_i \bar{\mu}_i + \varepsilon$, where ε denotes the experimental noise and approximation error. Consequently, for m instances of time $t_1, \ldots, t_m$, we obtain a system of equations $Y = AM + \varepsilon$ with

$$y(t_1) = A_1 M + \varepsilon_1 \tag{3.32}$$

$$y(t_2) = A_2 M + \varepsilon_2 \tag{3.33}$$

$$\vdots$$

$$y(t_m) = A_m M + \varepsilon_m. \tag{3.34}$$

Now, if Y represents a vector of experimental datasets on the outcome of a biological process with hysteresis behavior, then the identification problem for the density function μ is reduced to the minimization of the function

$$f(M) = \frac{1}{2} \| AM - Y \|^2 \tag{3.35}$$

with $A : \mathbb{R}^n \rightarrow \mathbb{R}^m$ $(Rank(A) = m)$, $M \in \mathbb{R}^n$ and $Y \in \mathbb{R}^m$. The Lagrange multiplier theorem provides the necessary conditions for the minimization by the existence of a $\lambda \in \mathbb{R}^n$ such that

- $\lambda_i = 0$ if $\mu_i \neq 0$;
- $\lambda_i \geq 0$ if $\mu_i = 0$; and
- $\bar{f}(M) = f(M) - \lambda^T M$ satisfies $\frac{\partial \bar{f}(M)}{\partial M} = M^T A^T A - Y^T A - \lambda^T = 0$.

The main advantage of this method is its good ability in identifying Preisach functions and thereby hysteresis systems without definition any prior knowledge of the density function. However, the appropriateness of this method for specific applications depends on the computational costs of the optimization process, which should be analyzed case-dependently.

References

Barker JA, Schreiber DE, Huth BG, Everett DH (1985) Magnetic hysteresis and minor loops—models and experiments. Proc R Soc London A 386:251–261

Bate G (1962) Statistical stability of preisach diagram for particles of γ-Fe_2O_3. J Appl Phys 33:2263–2269

Biorci G, Pescetti D (1958) Analytic theory of the behaviour of ferromagnetic materials. Nuovo Cinento 7:829–842

Biorci G, Pescetti D (1959) Some consequences of the analytical theory of the ferromagnetic hysteresis. J Phys Radium 20:233–236

Biorci G, Pescetti D (1966) Some remarks on hysteresis. J Appl Phys 37:425–427

Brokate M, Visintin A (1989) Properties of the preisach model for hysteresis. J Reine Angew Math 402:1–40

Brown WF Jr (1962) Failure of local-field concept for hysteresis calculations. J Appl Phys 33:1308–1309

Damlanian A, Visintin A (1983) A multidimensional generalization of the Preisach model for hysteresis. Compt Rend Acad Sci Paris 297:437–440

Torre Della E (1965) Measurements of interaction in an assembly of γ?iron oxide particles. J Appl Phys 36:518–522

Enderby JA (1956) The domain of hysteresis. 2. Interacting domains. Trans Faraday Soc 52:106–120

Everett DH (1954) A general approach to hysteresis. 3. A formal treatment of the independent domain model. Trans Faraday Soc 50:1071–1096

Everett DH (1955) A general approach to hysteresis. 4. An alternative formulation of the domain model. Trans Faraday Soc 51:1551–1557

Everett DH, Whitton WI (1952) A general approach to hysteresis. Trans Faraday Soc 48:749–757

Krasnosel'skii MA, Pokrovskii AV (1989) Systems with hysteresis. Springer, Heidelberg

Mayergoyz ID (1986) Mathematical models of hysteresis. IEEE Trans Magn 22:603–608

Mayergoyz ID (2003) Mathematical models of hysteresis and their applications. Elsevier, New York

Preisach F (1935) Über die magnetische Nachwirkung. Z Phys 94:277–302

Shirley ME, Venkataraman R (2003) On the identication of preisach measures. Proc SPIE 5049, 326–336

Visintin A (1984) On the Preisach model for hysteresis. Nonlinear Anal 8:977–996

Woodward JG, Della Torre E (1960) Particle interaction in magnetic recording tapes. J Appl Phys 31:56–62

Part II
Hysteresis in Biology

Chapter 4
Examples of Hysteresis Phenomena in Biology

Hysteresis may occur in different spatiotemporal scales of consideration. From switches in protein-DNA interactions (Chatterjee et al. 2008), microscopic cellular signaling pathways with bistable molecular cascades (Angeli et al. 2004; Qiao 2007), cell division, differentiation, cancer onset and apoptosis (Sha et al. 2003; Eissing et al. 2004; kim et al. 2007; Wilhelm 2009), protein folding (Andrews et al. 2013) and purinergic neuron-astrocyte interactions in the brain (Noori 2011) up to macroscopic biomechanics of cornea (Congdon et al. 2006) and lung deformations (Escolar and Escolar 2004), hysteresis phenomena are ubiquitous in biology. In this chapter, we will briefly discuss a few examples of hysteresis phenomena in biology, which may qualitatively resemble many other biological processes.

4.1 Hysteresis in Cell Biology and Genetics

4.1.1 Cell Cycle and Apoptosis

The dynamics of the eukaryotic cell cycle is shown to be governed by sequential activation and inactivation of cyclin-dependent protein kinases (CDKs (Pines 1995; Harper et al. 2002)). The CDK for entry into mitosis is Cdc2, whose activation requires binding to cyclin B (a regulatory protein) and the activation of phosphorylation process. However, both processes appears to provide only necessary conditions for the activation of Cdc2, as it can still be inactivated by inhibitory phosphorylations, which are carried out by protein kinases Wee1 and Myt1. Cdc25 protein phosphatases (as the immediate triggers for entry into mitosis) remove these Inhibitory phosphorylations and in their presence, the conditions become also sufficient for Cdc2 activation. CDKs regulating further cell cycle transitions can also be inhibited by direct binding of inhibitory proteins. Mitotic cyclins are subject to ubiquitin-mediated degradation at the end of mitosis by the action of the anaphase-promoting complex. In addition, the proper functioning of the cell cycle relies on

H. R. Noori, *Hysteresis Phenomena in Biology*, SpringerBriefs in Mathematical Methods, DOI: 10.1007/978-3-642-38218-5_4, © The Author(s) 2014

control mechanisms ensuring the correct sequence of key cell cycle events. For instance, specific DNA replication checkpoint ensures that Cdc2 is not activated until DNA replication is complete and the spindle assembly checkpoint acting as a switch prevents cyclin degradation until all chromosomes form the proper alignment on the metaphase plate (Solomon 2003).

By crushing frog eggs in the presence of minimal amounts of buffer one can obtain the Xenopus egg extracts that have lead to major advancements in understanding entry into and exit from mitosis (Pines 1995). Xenopus egg extracts are able to undergo multiple rapid cell cycles, due to either the morphology of added nuclei or by assays of Cdc2 activity. These cell cycles can be driven by the endogenous synthesis and degradation of cyclin, and in case of inhibited protein synthesis, by the addition and subsequent degradation of exogenous cyclin (Solomon 2003). The specific properties of this system, such as the lack of checkpoints and nonessential features one associates with more complex cell cycles in tissue culture as well as it ability to exhibit feedback loop, makes it an almost ideal system for biochemical investigation of cell cycle dynamics.

Using a mathematical model of cell-cycle progression in cell-free Xenopus egg extracts Sha et al. (2003) have shown that irreversible transitions into and out of mitosis are governed by hysteresis in the molecular control system. The mathematical model further identified the positive feedback in the phosphorylation reactions controlling the activity of Cdc2 as the underlying mechanism for the generation of the hysteresis behavior. This study is a great example of the research strategy that the present book attempts to facilitates, namely providing insights in biology by using mathematical models of hysteresis, which in this case enlightened specific mechanisms of the cell cycle dynamics.

However, the involvement of hysteresis in cellular processes is not restricted to mitosis, but is present during different developmental stages of a cell until apoptosis.

Apoptosis refers to a genetically determined form of programmed cell death utilizing the organism with a removal mechanism of unwanted cells at different times such as during embryonal development or after immune responses, to select educated immune cells and to eliminate virally infected and transformed cells (Leist and Jaattela 2001; Hengartner 2000). The human body reacts sensitively to alterations of apoptosis process and responds to variations in the apoptotic cell death rate severely through pathophysiological reactions in form of different diseases such as developmental defects, autoimmune diseases, neurodegeneration, or cancer. The apoptotic signaling pathways can be classified into two groups of intrinsic and extrinsic pathways, which partially share common signal transduction pathways. A significant sign of the ongoing apoptotic process is the activation of caspases, a family of aspartate-directed cysteine proteases that are produced as proenzymes.

The activation of caspases (upon cleavage of regulatory and structural proteins (Thornberry and Lazebnik 1998)) abolish the cells and induce the removal of dying cells by phagocytes.

An intrinsic feature of the apoptosis process is bistability. It is obvious as the status alive must be stable and resistant toward minor accidental trigger signals (Tyson et al. 2003). This behavior has already been shown experimentally and theoretically for several signal transduction pathways governing apoptotic processes (Xiong and Ferrell 2003). In this study, the authors assume the maturation of Xenopus oocytes as a process of cell fate induction and investigate the positive feedback loops of the p42 mitogen-activated protein kinase (MAPK) and the cell-division cycle protein kinase Cdc2. In general, such positive feedback loops could not only generate bistable behavior but furthermore create an non-local 'memory' of a transient inductive stimulus and could explain the irreversibility of maturation.

It appears that cell cycle and fate are governed by non-linear processes of hysteresis type, which have been partially analyzed by mathematical modeling and simulation. The success of these models in improving the understanding of underlying biological processes shows the power of mathematical models of hysteresis to analyze complex biological processes.

4.1.2 Synthetic Gene Networks

Synthetic biology is a scientific discipline that uses controllable, synthetic genetic devices to establish cells and organisms with predictable properties. The designed devices are either replacements of natural pathways or in combination with endogenous pathways generate complex artificial circuits. Although the creation of such synthetic gene circuits is a difficult procedure in general, previous studies have already demonstrated its feasibility even for mammalian systems paving the way for applications of synthetic networks in pharmacologic gene therapy (Weber and Fussenegger 2006; May et al. 2008). These pioneering investigations designed devices that expressed toggle and hysteretic switch behavior (kramer et al. 2004; Kramer and Fussenegger 2005).

In general, positive feedback control mechanisms in synthetic networks are responsible for the generation of binary states of either bistable switch (Becskei et al. 2001; Isaacs et al. 2003) or hysteresis (Xiong and Ferrell 2003; Angeli et al. 2004) type. The bistable expressions resemble the behavior of hysterons and jump from the down to up states and conversely, as a consequence of the input function reaching the threshold values α and β. Networks expressing hysteresis require larger signals for the transition between the states and depend on the integration of current as well as historic input signals.

As discussed by Kramer and Fussenegger (2005), the epigenetic toggle switch reveals probably the most extreme form of hysteresis as it independent of the presence of switch-triggering signals achieves the two stable expression states. Endogenous and synthetic multi-stable expression systems have been thoroughly examined particularly in Escherichia coli bacteria (Ozbudak et al. 2004; Atkinson et al. 2003) showing that positive feedback alone was insufficient for multi-stability. Similar to systems without any positive feedback loop mechanisms, hysteretic networks with imbalanced expression of network components produced a graded dose-response (Ozbudak et al. 2004).

The eminent presence of mathematical modeling to analyze hysteretic expression patterns within the synthetic gene networks shows the awareness of researchers in this field of research on the importance of quantitative understanding of this phenomenon. Recent studies (May et al. 2008) even utilized mathematical modeling to predict the hysteresis patterns created by synthetic positive feedback loops and verified their predictions experimentally. The authors have been able to show that the hysteretic response of feedback modules can be modulated by the strength of the positive feedback suggesting that for the establishment of complex multi-gene networks, several non-interacting switches (transactivators) or relais (promoters) are needed. The mathematical formalism for the investigation of such systems is provided by the Preisach models of hysteresis that has been introduced in part I of this monograph.

4.2 Hysteresis in Neuroscience

4.2.1 Puringeric Interactions of Neurons and Astrocytes

The nervous system consists of two cell types, the neuron and the glia. Recent experimental and theoretical studies have demonstrated that there is exists a dynamical reciprocal feedback relationship between neurons and perisynaptic glia cells (Araque et al. 1999; Haydon 2001; Newman 2003b; Volterra and Meldolesi 2005; Noori 2011). The release of neuronal neurotransmitters evoke an enhancement of Ca^{2+} concentrations in perisynaptic glia cells by binding to a wide variety of neurotransmitter receptors (Porter and McCarthy 1997). Consequently, the activated glia cells release various neuroactive substances in return, including glutamate and ATP. These transmitters influence the surrounding neurons by different routes of action: (1) they may feed back onto the presynaptic terminal either to facilitate or to suppress further release of neurotransmitter, and (2) they may stimulate directly the postsynaptic neurons by producing either excitatory or inhibitory responses (Fields and Stevens 2000; Halassa et al. 2007; Hamilton et al. 2008; Pascual et al. 2005; Haydon and Carmignoto 2006; Haydon et al. 2009; Perea 2005; Zhang, et al. 2003).

Glutamate released by astrocytes arouse slow inward currents through activation of postsynaptic NMDA receptors (Perea 2005; Perea et al. 2009; Araque et al. 1998; Angulo et al. 2004). In addition, astrocytic glutamate might also autoreceptors localized at presynaptic terminals. Through activation of group I metabotropic glutamate receptors (mGluRs) (Perea and Araque 2007) or NMDA receptors (Jourdain et al. 2007) astrocytes enhance the frequency of spontaneous and evoked excitatory synaptic currents. Furthermore, released Adenosine triphosphate (ATP) from glia could excite neurons directly through activation of $P2X$-receptors. Alternatively, released ATP, once converted to adenosine, could inhibit neurons by activating A_1-receptors, as it does in the retina. Indeed, adenosine could act presynaptically via A_1- and A_2-receptors to either depress or potentiate synaptic transmission (Dunwiddie et al. 1997; Gordon et al. 2005; Halassa et al. 2007; Haydon and

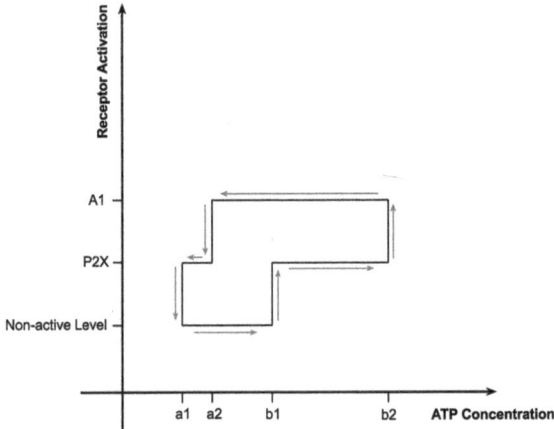

Fig. 4.1 The hysteresis diagramm of the ATP concentration and purinergic receptor activation. Two hysterons are parallel connected: First hysteron R_{a_1,b_1} is the delay relay operator of the accumulated to adenosine hydrolized ATP concentration; The second hysteron R_{a_2,b_2}, is the delay relay operator of the $P2$ activating ATP concentrations. The extremal values are denoting the concentration threshold of the release point of ATP and the maximum concentration of tripartite synaptic ATP. With author's permission, this figure is adapted from Noori (2011)

Carmignoto 2006; Kato et al. 2004; Martin et al. 2007; Masino et al. 2002; Newman 2003a, b). This process could easily be identified as a hysteresis phenomenon.

For the formulation of the ATP kinetics at synapses in the language of hysteresis, the relationship between ATP concentrations and activity behavior should be established. ATP can be released in concentrations that might be subthreshold for the activation of $P2X$-receptors (micromolar) yet sufficiently high to allow an accumulation of adenosine that will activate A_1-receptors (tens of nanomolar). This means, that the ATP kinetics can be decomposed into two phases (Fig. 4.1):

1. First Phase described by hysteron $h^1 = h_{a_1,b_1}$ of ATP concentrations acting on $P2X$-receptors. a_1 denotes the releasing point of ATP (concentration zero), while b_1 represents the ATP concentration activating $P2$ purinergic receptors.
2. Second Phase described by hysteron $h^2 = h_{a_2,b_2}$ of accumulated adenosine (hydrolized ATP) that acts inhibitory via A_1-receptor activation. a_2 denotes the maximum concentration level of ATP during pre-accumulation time of adenosine. b_2 is the accumulated adenosine concentration that is sufficient to activate A_1-receptors. Here, the hydrolization of ATP to adenosine is seen as a change in the activity state of ATP;

The parallel composition of these hysterons with equal weights μ_j leads to a discontinuous Preisach model of ATP hysteresis:

$$y(t) = \sum_{j=1}^{2} \mu_j h^j [t_0, \eta_0] x(t) \qquad (4.1)$$

where $y(t)$ denotes the activity induced by the concentration value of astrocytic ATP $x(t)$ and η_0 the initial state of the system.

In summary, the mechanisms of purinergic interactions of neurons and perisynaptic glia cells can be approximated by a simple discontinuous Preisach operator, which in turn enables us to adapt this model into further differential equations describing the non-linear neurochemical processes at synapses.

4.3 Hysteresis in Human Physiology and Anatomy

4.3.1 Cornea

The cornea is the primary infectious and structural barrier of the eye, consists of a transparent avascular connective tissue and in combination with the overlying tear film, it also provides a proper anterior refractive surface for the eye (DelMonte and Kim 2011). Through its shape (prolated flatter in the periphery and steeper centrally), cornea creates an aspheric optical system. In average, corneal horizontal diameter is between 11.5 to 12.0 mm (Rüfer et al. 2005) and 1.0 mm larger than the vertical diameter. The thickness of cornea at its center is approximately 0.5 mm, which gradually increases towards periphery.

Corneal form and curvature are governed by its morphologically defined intrinsic biomechanical properties and the environment. In particular, the Anterior corneal stromal rigidity appears to be crucial for the maintenance of the curvature of the cornea (Muller et al. 2001). Furthermore, differences in the organization of collagen bundles of the anterior stroma have been hypothesized to have an impact on the cohesive strength in this area and may be the cause of a higher resistance towards anterior curvature alterations with respect to stromal hydration than the posterior stroma (DelMonte and Kim 2011). Moreover, stromal hydration also appears to affect the corneal response to strain and shear forces (Simon and Ren 1994).

The cornea is not only the first defensive line of the eye against extrinsic factors, but corneal parameters, particularly central corneal thickness (CCT), also reflects intrinsic abnormalities and can be considered as potential determinants of both measured intraocular pressure (IOP) and glaucoma risk. To date, the most significant indication as well as the sole successful target for treatment of glaucoma is the intraocular pressure (Congdon et al. 2006), which can be obtained using different tonometry techniques.

In general, tonometry estimates the intraocular pressure and the so-called corneal hysteresis during rapid motion of the cornea in response to the shortduration (20 ms) air impulse, which causes an inwards movement (applanation) of cornea and a concave deformation of its shape. Subsequently, the air pump turns off and the cornea moves through a second applanation while reshaping from concavity to its original convex curvature (Fig. 4.2).

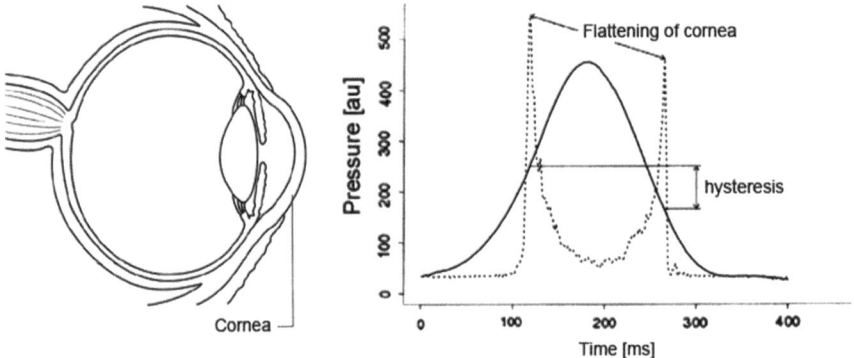

Fig. 4.2 Hysteresis behavior of human cornea using non-contact tonometer assessing the intraocular pressure. The *solid* and *dashed curves* represent the air pressure and light signal respectively. The peak of the Gaussian curve denotes the interface of in- and outward applanation of the air-puff pressure

The viscoelastic properties of the corneoscleral shell are the cause of the difference of the two applanation event pressures. The high speed of the corneal deformation induces velocity (rate)-dependent forces that oppose the pressure created by the air impulse, which by absorbing the energy from the air impulse create time delays (hysteresis behavior) in the occurrence of the applanation events (Congdon et al. 2006). The inward and outward applanation event pressures increase/decrease as a consequence of these delays. Therefore, the hysteresis behavior reflects an intrinsic viscoelastic biomechanical property of the cornea.

4.3.2 Respiratory System

The active component of the breathing process is called inspiration, which is initiated by the brain stem causing a contraction of the diaphragm (dome-shaped muscle separating the thoracic and abdominal cavities) and intercostal muscles and consequently an expansion of thoracic cavity and a decrease in the pleural space pressure. Due to elastic properties of the lungs, this process is passively reversed (expiration). However, the lungs dissipate energy and their deformation dynamics from one saturation point to the other cannot be described by a function but a functional. In other words, the applied energy to the lungs is not recovered completely by expiration process. This behavior is best observed by investigating the pressure-volume (P-V) relationship.

By placing the excised lung of an animal (commonly cats or dogs) in a pressure chamber, alterations of the lung volume can be measured with a spirometer through a cannula attached to the trachea. Decreasing the chamber pressure induces an increase of the lung's volume, whereas the gradual enhancement of the pressure leads to a decrease in the volume of the lungs (Fig. 4.3). It is noticeable that there exists a

Fig. 4.3 The respiratory pressure-volume loop formed by air inspiration (ascending) and expiration (descending) caused by the lung's energy dissipation (hysteresis)

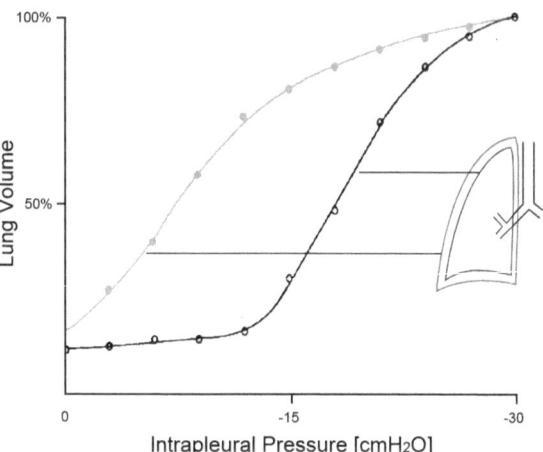

marked difference in the pressure change at the same volume, both for inflation and deflation (Harris 2005).

The area between the ascending and descending P-V curves is usually called lung hysteresis and refers to the above mentioned unrecoverable energy, or delayed recovery of energy, that is applied to the system. Lung hysteresis is two-fold of nature and consists of parenchymal and bronchial hysteresis. Each point on the P-V curve is associated with a different morphology of the lung parenchyma (Escolar and Escolar 2004).

Despite the fact that different hypotheses attempt to identify the forces, which by opposing the lung distension induce the hysteresis behavior, there is still no consensus on this issue. Surface forces have been hypothesized to be the predominant cause of hysteresis with large tidal excursions, however hysteresis with small volume excursions appears to be unrelated to these forces and is most likely caused by intrinsic tissue properties (Wilson 1981).

Various mathematical studies have already investigated the biomechanical properties of the lung as an elastic continuum undergoing small distortions from a state of uniform inflation (Mead et al. 1970; West and Matthews 1972; Bar-Yishay et al. 1986; Liu et al. 1990; Ganesan et al. 1995; Lai-Fook and Hyatt 2000). However, the uncertainty on the causes of hysteresis phenomena in respiratory system shows that the potential of mathematical modeling, particularly model for hysteresis non-linearities, is not yet exhausted and may provide significant contribution to this field of research.

References

Andrews BT, Capraro DT, Sulkowska JI, Onuchic JN, Jennings PA (2013) Hysteresis as a marker for complex, overlapping landscapes in proteins. J Phys Chem Lett 4:180–188

Angeli D, Ferrell Jr JE, Sontag ED (2004) Detection of multistability, bifurcations, and hysteresis in a large class of biological positive-feedback systems. Proc Nat Acad Sci U S A 101:1822–1827

Angulo MC, Kozlov AS, Charpak S, Audinat E (2004) Glutamate released from glial cells synchronizes neuronal activity in the hippocampus. J Neurosci 24:69206927

Araque A, Sanzgiri RP, Parpura V, Haydon PG (1998) Calcium elevation in astrocytes causes an NMDA receptor-dependent increase in the frequency of miniature synaptic currents in cultured hippocampal neurons. J Neurosci 18:68226829

Araque A, Parpura V, Sanzgiri RP, Haydon PG (1999) Tripartite synapses: glia, the unacknowledged partner. Trends Neurosci 22:208–215

Atkinson MR, Savageau MA, Myers JT, Ninfa AJ (2003) Development of genetic circuitry exhibiting toggle switch or oscillatory behavior in Escherichia coli. Cell 113:597–607

Bar-Yishay E, Hyatt RE, Rodarte JR (1986) Effect of heart weight on distribution of lung surface pressures in vertical dogs. J Appl Physiol 61:712–718

Becskei A, Seraphin B, Serrano L (2001) Positive feedback in eukaryotic gene networks: cell differentiation by graded to binary response conversion. EMBO J 20:2528–2535

Chatterjee A, Kaznessis YN, Hu WS (2008) Tweaking biological switches through a better understanding of bistability behavior. Curr Opin Biotechnol 19:475–481

Congdon NG, Broman AT, Bandeen-Roche K, Grover D, Quigley HA (2006) Central corneal thickness and corneal hysteresis associated with glaucoma damage. Am J Ophthalmol 141: 868–875

DelMonte DW, Kim T (2011) Anatomy and physiology of the cornea. J Cataract Refract Surg 37:588–598

Dunwiddie TV, Diao LH, Proctor WR (1997) Adenine nucleotides undergo rapid, quantitative conversion to adenosine in the extracellular space in rat hippocampus. J Neurosci 17:7673–7682

Eissing T, Conzelmann H, Gilles ED, Allgoewer F, Bullinger E, Scheurich P (2004) Bistability analyses of a caspase activation model for receptor-induced apoptosis. J Biol Chem 279:3689236897

Escolar JD, Escolar A (2004) Lung hysteresis: a morphological view. Histol Histopathol 19:159–166

Fields RD, Stevens B (2000) ATP: an extracellular signaling molecule between neurons and glia. Trends Neurosci 23:625–633

Ganesan S, Rouch KE, Lai-Fook SJ (1995) A finite element analysis of the effects of the abdomen on regional lung expansion. Respir Physiol 99:341–353

Gordon GRJ, Baimoukhametova DV, Hewitt SA, Kosala WRA, Rajapaksha JS, Fisher TE, Bains JS (2005) Norepinephrine triggers release of glial ATP to increase postsynaptic efficacy. Nat Neurosci 8:1078–1086

Halassa MM, Fellin T, Haydon PG (2007) The tripartite synapse: roles for gliotransmission in health and disease. Trends Mol Med 13:54–63

Hamilton N, Vayro S, Kirchhoff F, Verkhratsky A, Robbins J, Gorecki DC, Butt AM (2008) Mechanisms of ATP- and glutamate-mediated calcium signaling in white matter astrocytes. GLIA 56:734–749

Harper JW, Burton JL, Solomon MJ (2002) The anaphase-promoting complex: it's not just for mitosis any more. Genes and Dev 16:2179–2206

Harris RS (2005) Pressure-volume curves of the respiratory system. Resp Care 50:78–98

Haydon PG (2001) GLIA: listening and talking to the synapse. Nat Rev Neurosci 2:185–193

Haydon PG, Carmignoto G (2006) Astrocyte control of synaptic transmission and neurovascular coupling. Physiol Rev 86:1009–1031

Haydon PG, Blendy J, Moss SJ, Jackson FR (2009) Astrocytic control of synaptic transmission and plasticity: a target for drugs of abuse? Neuropharmacology 56:83–90

Hengartner MO (2000) The biochemistry of apoptosis. Nature 407:770–776

Isaacs FJ, Hasty J, Cantor CR, Collins JJ (2003) Prediction and measurement of an autoregulatory genetic module. Proc Nat Acad Sci USA 100:7714–7719

Jourdain P, Bergersen LH, Bhaukaurally K, Bezzi P, Santello M, Domercq M, Matute C, Tonello F, Gundersen V, Volterra A (2007) Glutamate exocytosis from astrocytes controls synaptic strength. Nat Neurosci 10:331339

Kato F, Kawamura M, Shigetomi E, Tanaka J-I, Inoue K (2004) ATP- and adenosine-mediated signaling in the central nervous system: synaptic purinoceptors: the stage for ATP to lay its "dual-role". J Pharmaco Sci 94:107–111

Kim D, Rath O, Kolch W, Cho K-H (2007) A hidden oncogenic positive feedback loop caused by crosstalk between Wnt and ERK pathways. Oncogene 26:45714579

Kramer BP, Viretta AU, Daoud-El-Baba M, Aubel D, Weber W, Fussenegger M (2004) An engineered epigenetic transgene switch in mammalian cells. Nat Biotechnol 22:867–870

Kramer BP, Fussenegger M (2005) Hysteresis in a synthetic mammalian gene network. Proc Nat Acad Sci USA 102:9517–9522

Lai-Fook SJ, Hyatt RE (2000) Effects of age on elastic moduli of human lungs. J Appl Physiol 89:163–168

Leist M, Jaattela M (2001) Four deaths and a funeral: from caspases to alternative mechanisms. Nat Rev Mol Cell Biol 2:589–598

Liu S, Margulies SS, Wilson TA (1990) Deformation of the dog lung in the chest wall. J Appl Physiol 68:1979–1987

Martin ED, Fernandez M, Perea G, Pascual O, Haydon PG, Araque A, Cena V (2007) Adenosine released by astrocytes contributes to hypoxia-induced modulation of synaptic transmission. GLIA 55:36–45

Masino SA, Diao L, Illes P, Zahniser NR, Larson GA, Johansson B, Fredholm BB, Dunwiddie TV (2002) Modulation of Hippocampal Glutamatergic Transmission by ATP Is Dependent on Adenosine A1 Receptors. J Pharm Exp Ther 303:356–363

May T, Eccleston L, Herrmann S, Hauser H, Goncalves J, Wirth D (2008) Bimodal and hysteretic expression in mammalian cells from a synthetic gene circuit. PLoS One 3:e2372

Mead J, Takishima T, Leith D (1970) Stress distribution in lungs: a model of pulmonary elasticity. J Appl Physiol 28:596–608

Muller LJ, Pels E, Vrensen GFJM (2001) The specific architecture of the anterior stroma accounts for maintenance of corneal curvature. Br J Ophthalmol 85:437443

Newman EA (2003a) Glial cell inhibition of neurons by release of ATP. J Neurosci 23:1659–1666

Newman EA (2003b) New roles for astrocytes: regulation of synaptic transmission. Trends Neurosci 26:536–542

Noori HR (2011) Substantial changes in synaptic firing frequencies induced by glial ATP hysteresis. Biosystems 105:238–242

Ozbudak EM, Thattai M, Lim HN, Shraiman BI, Van Oudenaarden A (2004) Multistability in the lactose utilization network of Escherichia coli. Nature 427:737–740

Pascual O, Casper KB, Kubera C, Zhang J, Revilla-Sanchez R, Sul J-Y, Takano H, Moss SJ, McCarthy K, Haydon PG (2005) Astrocytic purinergic signaling coordinates synaptic networks. Science 310:113–116

Perea G, Araque A (2005) Properties of synaptically evoked astrocyte calcium signal reveal synaptic information processing by astrocytes. J Neurosci 25:21922203

Perea G, Araque A (2007) Astrocytes potentiate transmitter release at single hippocampal synapses. Science 317:10831086

Perea G, Navarrete M, Araque A (2009) Tripartite synapses: astrocytes process and control synaptic information. Trends Neurosci 32:421–431

Pines J (1995) Cyclins and cyclin-dependent kinases: a biochemical view. Biochem J 308:697711

Porter JT, McCarthy KD (1997) Astrocytic neurotransmitter receptors in situ and in vivo. Prog Neurobiol 51:439–455

Qiao L, Nachbar RB, Kevrekidis IG, Shvartsman SY (2007) Bistability and oscillations in the Huang-Ferrell model of MAPK signaling. PLoS Comput Biol 3:1819–1826

Rüfer F, Schrder A, Erb C (2005) White-to-white corneal diameter; normal values in healthy humans obtained with the Orbscan II topography system. Cornea 24:259261

Sha W, Moore J, Chen K, Lassaletta AD, Yi C-S, Tyson JJ, Sible JC (2003) Hysteresis drives cell-cycle transitions in Xenopus laevis egg extracts. Proc Nat Acad Sci USA 100:975–980

Simon G, Ren Q (1994) Biomechanical behavior of the cornea and its response to radial keratotomy. J Refract Corneal Surg 10:343351

Solomon MJ (2003) Hysteresis meets the cell cycle. Proc Nat Acad Sci USA 100:771–772

Thornberry NA, Lazebnik Y (1998) Caspases: enemies within. Science 281:1312–1316

Tyson JJ, Chen KC, Novak B (2003) Sniffers, buzzers, toggles and blinkers: dynamics of regulatory and signaling pathways in the cell. Curr Opin Cell Biol 15:221231

Volterra A, Meldolesi J (2005) Astrocytes, from brain glue to communication elements: the revolution continues. Nat Rev Neurosci 6:626–640

Weber W, Fussenegger M (2006) Pharmacologic transgene control systems for gene therapy. J Gene Med 8:535–556

West JB, Matthews FL (1972) Stresses, strains, and surface pressures in the lung caused by its weight. J Appl Physiol 32:332–345

Wilhelm T (2009) The smallest chemical reaction system with bistability. BMC Syst Biol 3:90

Wilson TW (1981) Relations among recoil pressure, surface area and surface tension in the lung. J Appl Physiol Respirat Environ Exercise Physiol 50:921–926

Xiong W, Ferrell JE Jr (2003) A positive-feedback-based bistable 'memory module' that governs a cell fate decision. Nature 426:460–465

Zhang J-M, Wang H-K, Ye C-Q, Ge W, Chen Y, Jiang Z-L, Wu C-P, Poo M-M, Duan S (2003) ATP released by astrocytes mediates glutamatergic activity-dependent heterosynaptic suppression. Neuron 40:971–982